TECHNOLOGY: A SECOND LEVEL COURSE

T264 BLOCK 3

DESIGN

PRINCIPLES AND PRACTICE

BLOCK 3

CREATIVITY AND CONCEPTUAL DESIGN: THE INVENTION AND EVOLUTION OF BICYCLES

PREPARED FOR THE COURSE TEAM BY
ROBIN ROY

T264 DESIGN: PRINCIPLES AND PRACTICE

CONTENTS OF THE COURSE

BLOCK 1 **AN INTRODUCTION TO DESIGN**

BLOCK 2 **PRODUCT PLANNING AND THE DESIGN BRIEF**

BLOCK 3 **CREATIVITY AND CONCEPTUAL DESIGN**

BLOCK 4 **GEOMETRY AND CONFIGURATION IN DESIGN**

BLOCK 5 **PRODUCT DEVELOPMENT AND MANUFACTURE**

BLOCK 6 **A REVIEW OF DESIGN**

The Open University, Walton Hall, Milton Keynes, MK7 6AA.

First published 1992.

Edited, designed and typeset by the Open University.

Printed in the United Kingdom by BPCC Wheatons Ltd.

This text forms part of an Open University Second Level Course. If you would like a copy of *Studying with the Open University*, please write to the Central Enquiry Service, PO Box 200, The Open University, Walton Hall, Milton Keynes, MK7 6YZ, United Kingdom. If you have not already enrolled on the Course and would like to buy this or other Open University material, please write to Open University Educational Enterprises Ltd, 12 Cofferidge Close, Stony Stratford, Milton Keynes, MK11 1BY, United Kingdom.

ISBN 0 7492 6102 1

Edition 1.1

CONTENTS

	STUDY GUIDE	5
1	THE ROLE OF CREATIVITY IN INVENTION AND DESIGN	8
1.1	What is creativity?	8
1.2	Invention, design and innovation	10
1.3	Creativity and conceptual design	15
1.4	Different types of thinking involved in design	19
1.5	Audio: Types of thinking in design	22
2	INVENTION AND EVOLUTION OF THE BICYCLE	23
2.1	Invention of the pedal cycle	23
2.2	Nineteenth-century cycle evolution	25
2.3	Twentieth-century cycle evolution	27
2.4	Video: The evolution of the bicycle	31
3	CREATIVITY IN THEORY	33
3.1	The creative process	33
3.2	Associative theories of creativity	36
3.3	Where do inventions and design ideas come from?	37
3.4	Video: Creativity and innovation	45
4	CREATIVITY IN PRACTICE: NINETEENTH-CENTURY BICYCLE DESIGN	47
4.1	The Rover Safety bicycle	47
4.2	The pneumatic tyre	55
4.3	The Dursley Pedersen bicycle	58
5	THE INVENTOR AND CREATIVE DESIGNER	63
5.1	The creative personality	63
5.2	Creative abilities and skills	64
5.3	Knowledge and creativity	69
5.4	Adaptors and innovators	70
5.5	Audio: Thinking styles exercises	71
5.6	Creative design problems	71
6	CREATIVITY IN PRACTICE: TWENTIETH-CENTURY BICYCLE DESIGN	74
6.1	The Moulton bicycle	77
6.2	Mountain bicycles	88
6.3	Video: The challenge of the portable bike	89
7	APPROACHES AND TECHNIQUES FOR INVENTION AND CREATIVE DESIGN	93
7.1	Conventional approaches to invention and creative design	93
7.2	Creative Problem Solving techniques	102
7.3	Audio: Creativity techniques	103
7.4	Eight idea generation techniques	103
7.5	Evaluation and selection of ideas	107
7.6	Audio: Idea generation techniques	108
8	GOOD IDEAS ARE NOT ENOUGH	109
8.1	Barriers to innovation	109
8.2	Barriers to diffusion	111
8.3	Two examples	115
8.4	A final word	120

ANSWERS TO SELF-ASSESSMENT QUESTIONS 121

CHECKLIST OF OBJECTIVES 125

REFERENCES 128

AUTHOR'S ACKNOWLEDGEMENTS

I am grateful to Sally Boyle, Barry Dagger, Paul Gardiner, Veronika Roy, Reg Talbot and David Gordon Wilson for their helpful comments, advice and assistance during the preparation of this Block.

STUDY GUIDE

AIMS

The aims of this Block are:

- to focus on the inventive and creative aspects of designing throughout the design process that was outlined in Block 2, but especially in the early conceptual phase of design;
- to provide you with an understanding of the principles of creative thinking in design through an examination both of theories and of practical techniques;
- to help you apply these principles and techniques in tackling the Guided Design Exercise for this Block, in which alternative design ideas and conceptual solutions to a given product development problem are generated and evaluated;
- to show that inventiveness and creativity are necessary, but not sufficient, for new technical ideas and design concepts to become product innovations that survive in the market.

These ideas about invention, creativity and innovation are illustrated using mainly (but not exclusively) the case of bicycles and their components. The bicycle is an enormously successful product which has evolved into its present forms over a period of about 170 years, as a result of several important inventions and numerous design developments, and is still evolving today.

OUTCOMES

After you have studied this Block, you should:

- understand the role of creativity in invention and design;
- have learned about some of the main theories of creativity and be able to relate these to examples of creative thinking and techniques for generating design ideas;
- understand where inventive and creative designers get their ideas from and the mental processes involved in creative thinking;
- appreciate what skills and abilities are involved in creative design;
- be aware of your own style of thinking and to choose appropriate techniques for creative problem solving;
- be able to apply approaches and techniques to enhance your creative and inventive thinking for idea generation and conceptual design;
- be able to discuss why creativity and inventiveness are not enough to ensure the successful development and diffusion of new products or innovations.

A detailed Checklist of Objectives is given at the end of the Block.

WHAT YOU HAVE TO DO

Block 3 of this course comprises:

- this main Block text (Sections 1–8) with exercises, in-text and self-assessment questions;
- a video-cassette divided into three sections, together with the *Video 3 Study Guide*;
- an audio-cassette in five parts, together with the *Audio 2 Study Guide* which includes tuition in techniques for the Guided Design Exercise (GDE);
- the Guided Design Exercise for the Block;
- Tutor-marked Assignment T264 TMA 03.

The study chart opposite shows how the various components relate to each other.

Most of the information you will need to study the Block is included in this text. You will be asked at appropriate points to view the video and listen to the audio-cassette and to start thinking about the GDE. This does not mean that you have to study the Block in this precise order; feel free to dip into the various Block components in any way that catches your interest. But you will get the most out of the Block if you follow the recommended study pattern, at least *within* any particular section of the Block. So, for example, studying Section 2 includes watching Video 3, Section 1 and attempting some self-assessment questions. This is the best place at which to view that section of the video, although you may prefer to study, say, Section 3 of the text before Section 2.

As in the previous Block, throughout this text various short exercises occur, which are sometimes followed by notes. These exercises and the accompanying notes are printed in brown type. Although these exercises will take up relatively little of your time, some of them may not be possible for you to do exactly at the moment they occur in the text. In that case, *do not* read the notes immediately following the exercise until you have had the chance to try it for yourself – go on to the next piece of normal text. (Not all the exercises have notes to follow them – in those cases you should make your own appraisal of the exercise.)

You will see that the sections of this main Block text alternate between general and theoretical material on creativity in invention and conceptual design and specific case studies of innovative designing, mainly using the example of pedal cycles. But remember that the purpose of the case studies and examples is *not* to teach you about bicycles – or any other product – as such, but to help you understand the general principles of creative and inventive thinking. And, together with the other material, the examples are intended to help improve your ability to think creatively when faced with a product or engineering design problem. The TMA covers both theoretical and practical aspects of the Block. The GDE is concerned with the practical side and in tackling it you will be using some of the idea generation techniques introduced in Section 7 of this text. However, the text only *introduces* the techniques. You *learn* how to use them with the help of the audio-cassette and the accompanying *Audio 2 Study Guide* provided as supplementary material. The exercises in Sections 3, 5 and 7 are also intended to help you with the GDE, so it is worth spending the 10–15 minutes needed to attempt each of these.

So, before you start, get a feel for the main Block components by quickly looking through this text and the various supplementary materials, including TMA 03. The earlier you start thinking about the TMA the better because, as you will see from the Block, good preparation is an essential part of creative problem solving.

STUDY CHART

Section	Main text	Exercises	Video	Audio
1	The role of creativity in invention and design	Alarm clock concepts		Audio 2, Side 1, Part 1: Types of thinking in design
2	Invention and evolution of the bicycle		Video 3, Section 1: The evolution of the bicycle	
3	Creativity in theory	Associative thinking (GDE)	Video 3, Section 2: Creativity and innovation	
4	Creativity in practice: nineteenth-century bicycle design		Video 3, Section 1: The evolution of the bicycle	
5	The inventor and creative designer	Creative design problems		Audio 2, Side 1, Part 2: Thinking styles exercises
6	Creativity in practice: twentieth-century bicycle design	Portable transport	Video 3, Section 3: The challenge of the portable bike	
7	Approaches and techniques for invention and creative design	Checklists (GDE)		Audio 2, Side 1, Part 3: Creativity techniques Audio 2, Side 2 Parts 1 and 2: Idea generation techniques
8	Good ideas are not enough	Successful innovation	Video 3, Section 2: Creativity and innovation	
		Guided Design Exercise; TMA 03		

1 THE ROLE OF CREATIVITY IN INVENTION AND DESIGN

Design is often described as a creative activity, and designers are usually regarded as creative people. But what does being creative mean? Why is creativity necessary? Where do ideas for inventions and new designs come from? Are creative designers special, or can anyone produce creative design work? How useful are creative problem solving techniques such as brainstorming (some of which are outlined in Section 7) in design? And how important are creative and inventive thinking relative to other abilities and skills for design and innovation?

In this Block I shall be exploring these and similar questions. Study of this Block should not only improve your theoretical understanding of the role of creativity in design, but enable you to become more creative and inventive in your own thinking and to apply this thinking to practical design and product development problems.

1.1 WHAT IS CREATIVITY?

First of all, what is meant by creativity? This is not an easy question to answer, as Donald Mackinnon pointed out in an address to a Nobel conference on the subject:

> Many are the meanings of creativity. Perhaps for most it denotes the ability to bring something new into existence, while for others it is not an ability but the psychological processes by which new and valuable products are fashioned. For still others, creativity is not the process but the product [...] creativity properly carries all these meanings and many more besides.
>
> (Mackinnon, 1970, p. 19)

So creativity means different things to different people because it may be associated with:

- an idea or object that is novel and valuable in some way (a creative *product*);
- the process that produced the idea or object (the creative *process*);
- the ability of an individual to produce such ideas or objects (a creative *person*).

You may remember that in Block 1 designing was defined as a mental process. In this Block I shall also focus mainly on the mental-process interpretation of creativity, although, as you will see, practical actions are usually also involved in creative work. On this definition, creativity is the thinking and other processes that bring original (new, unusual, novel, unexpected) ideas and objects into being. Of course an idea or object may be highly original without being of any value either to the thinker or to anyone else. So most definitions also specify that the idea or object must be considered *appropriate* or *worthwhile* for the process to be creative. Thus Stein (1956) defines **creativity** as: 'that process which results in a novel work that is accepted as tenable or useful or satisfying by a group at some point in time'. Whether this just means the moment of inspiration, or 'flash of insight', when a new idea occurs in the mind of the creator or the more extended process by which problems are solved and objects are produced, is a question I shall leave until later. What is clear is that in this course we are not concerned with every sort of creative activity – that would include, say, writing music, directing a play, or managing a business – but with the design of new products.

So, before going any further, let us look at what some designers have said about the way they go about getting ideas and developing new products.

First Alex Moulton, a creative engineering designer (whose best-known work, the design of small-wheel bicycles, is described in Section 6 and shown on Video 3, Section 1, 'The evolution of the bicycle') on what was involved in developing the innovative Hydrolastic suspension system used on the BMC Mini and 1100/1300 series of cars:

> These things do not appear on the stage by a flash of insight – important as this is – but from years of experiment, observation and consequent modifications. Above all it involves countless hours of thought and discussion with the intimate few ... It is better for one mind only to guide each step of a development so that the project evolves towards the vision held in the mind of the designer.
>
> (Moulton, 1979, p. 39)

Next James Dyson on his approach. Dyson is a highly inventive product designer who talks on Video 3, Section 2, 'Creativity and innovation' about how he conceived and developed his novel 'Ballbarrow' wheelbarrow and 'Cyclone' vacuum cleaner:

> I start off with an idea for a technical innovation. Something on which I will be able to get patent protection ...
>
> I am a great believer in two things. Firstly ... select a finite area and it is amazing how quickly you can become expert in it. Secondly, I believe in painstaking and logical testing. In every one of our projects, the patentable and inventive parts have been happened upon during the course of practical testing.
>
> (Dyson, 1987, p. 17)

Third, Mark Sanders, an engineer turned industrial designer (who appears on Video 3, Section 3, 'The challenge of the portable bike') on how he conceived his novel folding bicycle:

> A long time was spent thinking about the problem and jotting down ideas at random ... The competitors all appeared to have been designed to fold as small as possible in all directions. This was questioned: why not design the folded form long and thin like a walking stick? This decision was confirmed by considering one of the most successful products of recent years, the folding baby buggy, which folds to form a long thin stick with wheels at one end ... Having decided on this folded form, the structure of the extended bicycle was concentrated upon.
> [...]
> A matrix was drawn up with selected structures arranged horizontally and possible drive arrangements arranged vertically. The matrix ensured all possible combinations of structure and drive arrangement were considered.
>
> (Sanders, 1985)

I think that you can see from the above accounts that, in areas of design ranging from car suspensions to bicycles, creativity involves having new ideas. They also indicate that it involves transforming those ideas, usually with the aid of drawings, models and prototypes, into a design for a practical object that will work and can be made.

Sometimes the new design will be novel enough to count as a patentable invention. So, before we look at the creative process in more detail, let us first consider invention and how it relates to design and to product innovation.

1.2 INVENTION, DESIGN AND INNOVATION

INVENTION AND INNOVATION

According to the Chartered Institute of Patent Agents an **invention** is 'simply something new, something which has not been thought of before and which is not obvious'. Most inventions are new machines, products or manufacturing processes and these can generally be granted the protection of a **patent** provided: (a) the invention has never before been made public anywhere in the world; (b) it would not be considered 'obvious' by someone knowledgeable in the field of the invention; (c) the invention is capable of practical application. An invention may start as an idea for something new, but to be patentable it must be described in enough detail to be judged capable of being made into a practical device. Thus, as you will see on Video 3, Section 2, James Dyson's novel vacuum cleaner started as the idea of using the principle of an industrial cyclone separator for a domestic cleaning machine, but he could not patent his invention until he had developed a form of cylone which could collect dust and dirt in a domestic-scale device. (If you are interested in inventions and patents, they are discussed in more detail in the Open University's third-level course on design and innovation.)

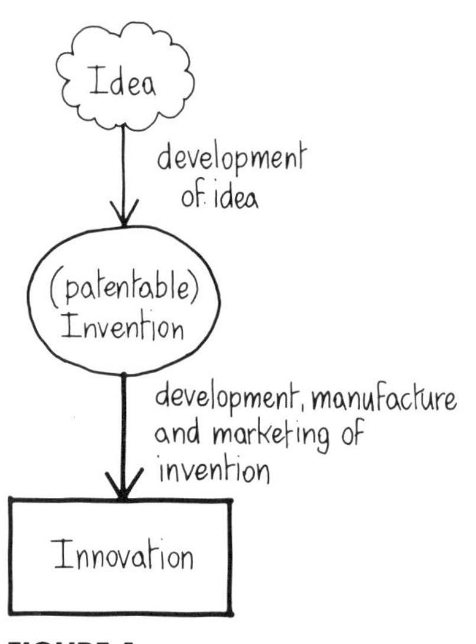

FIGURE 1
SIMPLE LINEAR MODEL OF THE INNOVATION PROCESS

What concerns us here is that only a minority of ideas and inventions lead to innovations. Only a small proportion of inventions, even patented ones, undergo development to the working prototype stage, and only a proportion of these prototypes are further developed to the production stage. And, even then, not all production designs are introduced on to the market or otherwise brought into use. Only at the point of *initial commercial introduction* (or adoption into use) is the invention said to have become an **innovation.** Figure 1 shows this process in a highly simplified diagrammatic form.

The distinction between an invention and an innovation was first made by an economist, Joseph Schumpeter, early this century. In this Block I am concerned with inventions of new machines and devices which result in *product* innovations, but the distinction between an invention and an innovation applies equally to new manufacturing processes, chemicals and materials. Innovation is also often used to describe the whole process by which a new idea or invention is created and developed into a commercial product, and it will be necessary sometimes to use the term in that sense too.

THE HOVERCRAFT AS A PRODUCT INNOVATION

Let me illustrate the complexity of the innovation process by outlining the development of a well-known product innovation – the hovercraft. The idea of a craft supported on a cushion of air was first proposed in the eighteenth century, and the first patent on the concept was granted in 1887, yet the hovercraft did not become a product innovation until the early 1960s.

The first working air-cushion vehicle was built by an Austrian engineer in 1915, but, despite further attempts to design a practical craft, none really succeeded because of the amount of power needed to maintain a stable, pressurised air cushion beneath the vehicle.

In the 1950s Christopher Cockerell began experimenting with air lubrication for boats, which led him to an idea for an improved type of craft supported on an air cushion enclosed in a space formed by pumping air in a 'curtain' round the vehicle's periphery (Figure 2). After verifying his idea, first with a test rig made from tin cans and an industrial blower (Figure 3), and then with a balsa wood model (Figure 4), Cockerell applied for a patent on his hovercraft concept in 1955 (Figure 5).

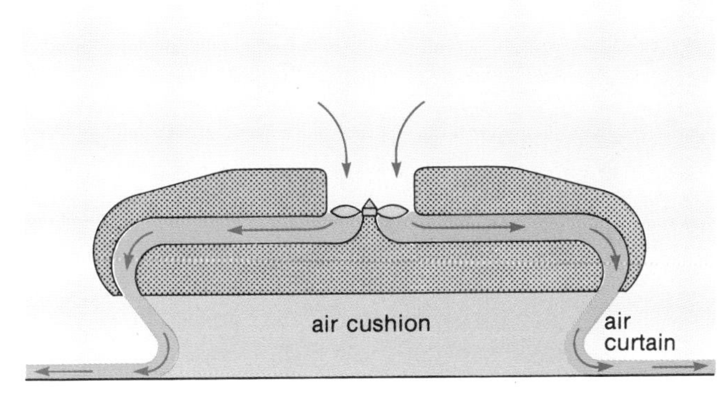

FIGURE 2
IDEA: COCKERELL'S PERIPHERAL JET HOVERCRAFT PRINCIPLE

FIGURE 3
VERIFYING IDEA: MOCK-UP FROM EMPTY TINS AND AN INDUSTRIAL BLOWER USED TO TEST THE HOVERCRAFT PRINCIPLE

FIGURE 4
INVENTION: CHRISTOPHER COCKERELL AND A.D. TRUEMAN WITH THE FIRST PROPER HOVERCRAFT MODEL MADE IN 1955 FROM BALSA WOOD

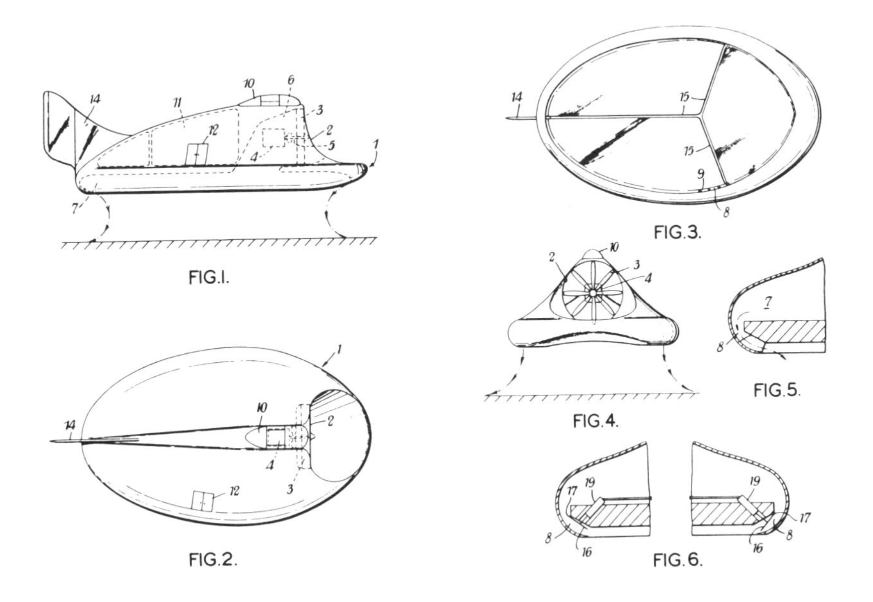

FIGURE 5
INVENTION: ONE SHEET OF THE DRAWINGS FROM COCKERELL'S PATENT SPECIFICATION

Cockerell then began the task of trying to get his invention developed. None of the established aircraft manufacturers or shipbuilders was interested, and Cockerell managed to obtain government support for the early development of the hovercraft only because of its military potential. In 1958, however, he managed to persuade the National Research Development Corporation to back the hovercraft, and by 1959 the first manned prototype vehicle employing Cockerell's invention had been designed and built. This was the SR.N1, built by Saunders Roe as a machine for research and development (Figure 6). It soon became apparent that the peripheral jet on the SR.N1 did not provide enough lift to make it a practical form of transport; this required the addition of a patented flexible 'skirt' to retain the air cushion. A series of experimental hovercraft designs for both civil and military applications followed the SR.N1, but it was not until the development of the SR.N5 and the SR.N6 (Figure 7), which were sufficiently attractive to travellers to allow the first commercial hovercraft service to come into operation in 1965, that the hovercraft could be considered as a product innovation.

FIGURE 6 (TOP)
PROTOTYPE DESIGN: THE WORLD'S FIRST MANNED HOVERCRAFT, THE PROTOTYPE SR.N1 USED FOR RESEARCH AND DEVELOPMENT

FIGURE 7
PRODUCT INNOVATION: ONE OF THE SR.N6 SERIES OF HOVERCRAFT. THE SR.N6 WAS USED IN COMMERCIAL SERVICE FROM THE MID 1960S TO THE EARLY 1980S

Even though the above account is highly simplified, you can see from this example that the road from an idea to the invention of a new machine or device to its introduction on to the market as a product innovation is often long and arduous. It is not surprising that most ideas and inventions never succeed in becoming innovations.

DIFFUSION AND EVOLUTION

An innovation of itself is rarely of much importance to society. For it to have any significant impact, the innovation has to be adopted by a large number of individuals, firms or organisations. In other words, it has to undergo **diffusion** into commercial or social use. If a product innovation is successful, it usually undergoes evolutionary design and development into improved versions, finds new applications, and may displace older products, devices and systems as it spreads into use.

Taking our hovercraft example again, the SR.N6 was followed in 1968 by the much larger SR.N4 and more than a dozen variants of these two designs were developed (Rothwell and Gardiner, 1985). In 1982 the AP-188, a new type of hovercraft, completely redesigned to halve production and running costs by using marine rather than aircraft technology, was introduced to replace the SR.N6. A number of other hovercraft designs were also developed for particular applications. In the hovercraft you can see that evolutionary design and development took over from basic invention and innovation.

Nevertheless, the evolution of a product frequently requires further invention and innovation. Usually this involves the creation of new components and materials, such as the synthetic rubbers used in hovercraft skirts. These are sometimes described as **incremental innovations** to distinguish them from the original **radical innovation.** You should recall examples of products evolving in this way from Blocks 1 and 2 and you will see the transition from radical innovation to incremental innovation and design variation even more clearly in the case of the bicycle, which is discussed in this Block.

As it diffuses and evolves, the machine or device ceases to be regarded as 'an innovation' any more and becomes simply one of the many familiar products available in society. Eventually the evolved product may decline and be displaced as further innovations come along. This appears to be happening with the hovercraft, which by the 1990s looked as if it was to be displaced by high speed catamarans for fast ferry services. In future, hovercraft may survive only as military or rescue vehicles.

Various attempts have been made to produce diagrams and flow charts that represent the various stages leading from the invention of a novel machine or device and its introduction as a product innovation to its diffusion and evolutionary development, but none is really satisfactory because of the complexity of the processes involved. My attempt to illustrate this complexity is shown in Figure 8. I have represented the process as a spiral because it tends to start with one or more individuals working on an idea or invention. If the invention is developed, increasing numbers of people and increasing resources become involved to produce a product innovation, and as the innovation diffuses many different designs based on the original idea may be generated. Note the importance of creative ideas at all stages, and that I have included a feedback loop from 'product evolution' back to invention. This is to indicate that, as the product is improved and redesigned, it may go through several further spirals of incremental innovation before it stabilises or declines.

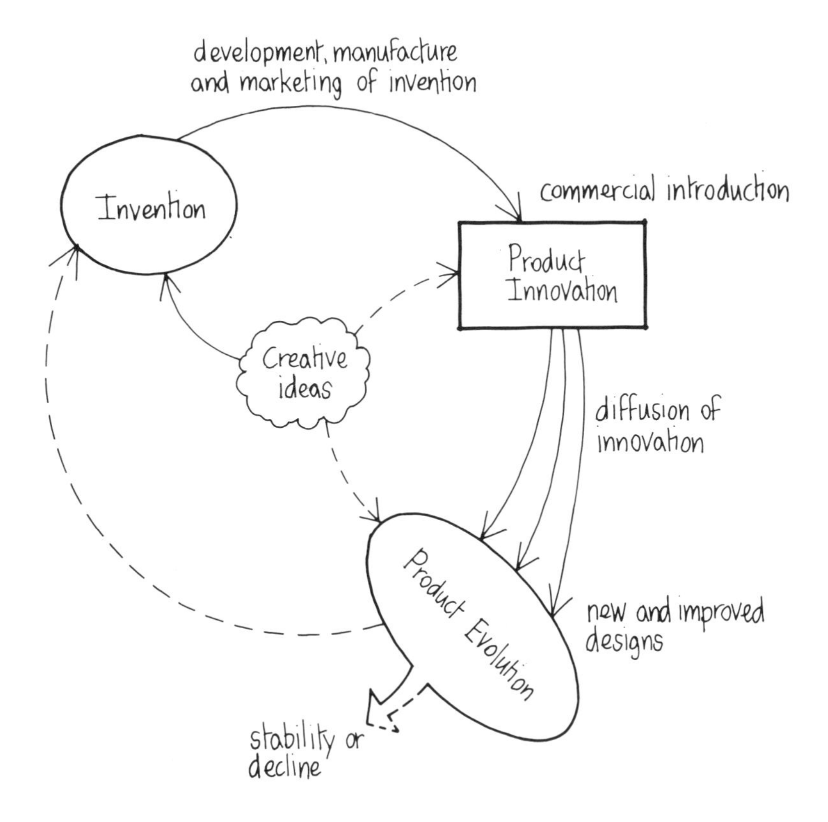

FIGURE 8
MODEL OF THE PROCESS OF PRODUCT INNOVATION, DIFFUSION AND EVOLUTION

DESIGN

How then does design fit into the processes I have just described?

Designing, like inventing, is a creative activity. But whereas invention involves proposing an advance in the known state-of-the-art of a particular field, most design involves making variations on the existing state-of-the-art. And, while invention can merely result in a sketch or model giving the principle of something new or improved, designing goes on to give the concept a specific physical form, or configuration.

Michael French notes:

> ... invention is nothing but a grander word for a particularly original or important step in [functional] design. Although invention is generally much more impressive than design, it often turns out to be the easiest part.
>
> A recent example in which invention was rather obvious, even trivial, and the difficulty lay in design, was the application of disc brakes to cars. The brakes on bicycles had been of this form for decades, but a great deal of ingenuity and hard work had to be put into producing a disc brake for cars that could compete with the highly-developed drum brake on favourable terms.
>
> ... In general, invention is more fundamental and less specific in form and application than design, but no sharp line can be drawn between the two.
>
> (French, 1988, p. 275)

Product innovation therefore involves both invention and design. Inventions describe a new idea in principle; designs work towards detailed prescriptions for making a practical device.

The gap between an invention in principle and a working design in practice may be considerable, and bridging the gap may take a long time. Many inventions, like the nineteenth-century hovercraft and, as you will see later, the pneumatic tyre, were years in advance of realisation as practical innovations because of the limitations of available materials, knowledge or technology.

You can see that a single invention can often give rise to a variety of different designs that utilise the same basic principle. In fact the great proportion of design work going on today is based on past inventions and innovations; only a small proportion of designing depends on new principles or new knowledge. So, although many experimental and prototype designs are often needed to bridge the gaps between an original idea, an invention and a product innovation, design can continue long after innovation has taken place, that is, during the stage of a product's evolutionary development. Designers rearrange components, devise new forms of assembly, take up the possibilities of new materials, and in this way make improvements to existing products, devices and systems. It is through design (and incremental innovation) that products evolve, vary and change.

Often an innovation becomes commercially or socially important only after its evolution into a variety of different designs for different markets and different applications. Product innovations, when first introduced, are often expensive, limited in performance and may be poor in design, for example being unreliable or difficult to use. Consequently innovations are often adopted only by enthusiasts or for essential industrial or military applications. Only through design evolution and technical improvement are such innovations made accessible and acceptable to a wide market. This was the case with the hovercraft, as you have just seen, and also the bicycle, as you will see later.

To sum up, invention and design are both creative activities because both are concerned with the origination of something that has not existed before; a novel principle in the case of invention and a new form of a product, device or system in the case of design. Product innovation requires both invention and design. Design, however, may continue long after the original innovation has been introduced.

SAQ 1

Using the example of the hovercraft, distinguish between the following terms: invention, design, innovation, diffusion.

1.3 CREATIVITY AND CONCEPTUAL DESIGN

So far I have not said much about where the original idea – or concept – for a new product comes from. This is one of the main topics of this Block and, as you will see later, product ideas or concepts can arise in a variety of ways. One important way is through the part of the design process called **conceptual design**.

By now you should know that the design of products, at least in an industrial context, typically proceeds through a number of broad stages as the object concerned is gradually defined more and more precisely until it is described in sufficient detail for it to be made. You may wish to refresh your memory of the total product design process by referring back to Block 2 (Section 12).

The conceptual phase is usually taken to be the early part of the process in which the major design decisions are taken. It is the phase during which the designer or design team takes the statement of the problem given in the brief or the requirements given in the product design specification (you should recall the meaning of these terms from Block 2) and creates broad outline solutions to the given problem or specified requirements. These solutions (or design concepts) are then evaluated to select, for further development, the concept which best satisfies the requirements of the specification.

The extent to which these design concepts define the final design varies depending on the problem and the area of design concerned. In some cases, the design concepts may be little more than rough sketches of initial ideas. In engineering, conceptual design usually goes beyond ideas in order to demonstrate technical feasibility, and the concepts are normally represented as drawings or diagrams which show how each major function is to be performed, together with the spatial and structural relationships of the major components. The importance of spatial and structural relationships in design is something you will be looking at in more detail in Block 4.

What is involved in conceptual design also depends very much on the starting point. Conceptual design may start at the level of a total product, such as a motor vehicle, or at the level of a particular sub-system, such as the vehicle's engine, or at the level of a component such as a water pump. If the starting point is the total product level, conceptual design may involve the generation of new ideas and inventions which if implemented would result in a major product innovation. At the component level it may be more like solving a detailed design problem.

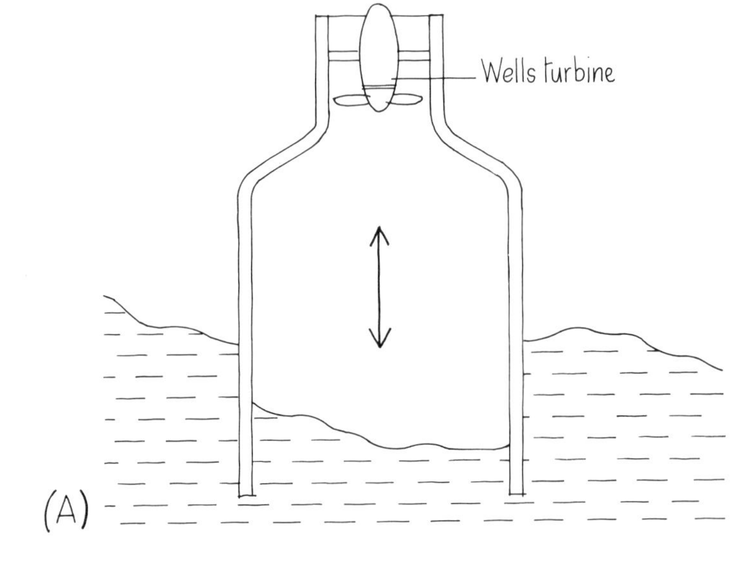

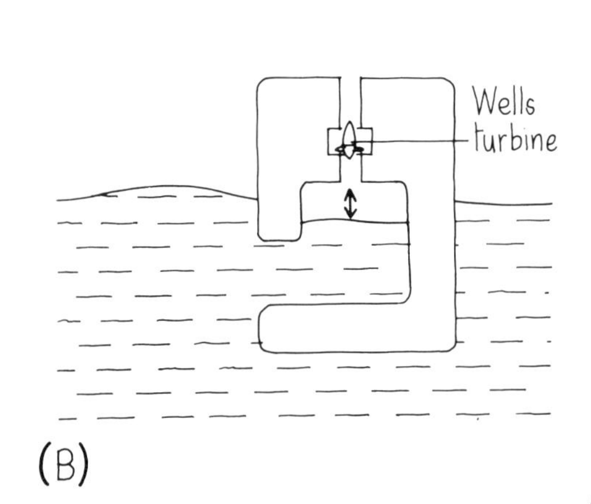

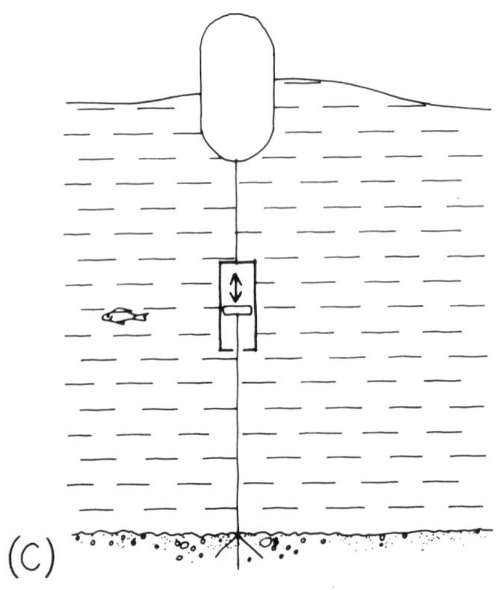

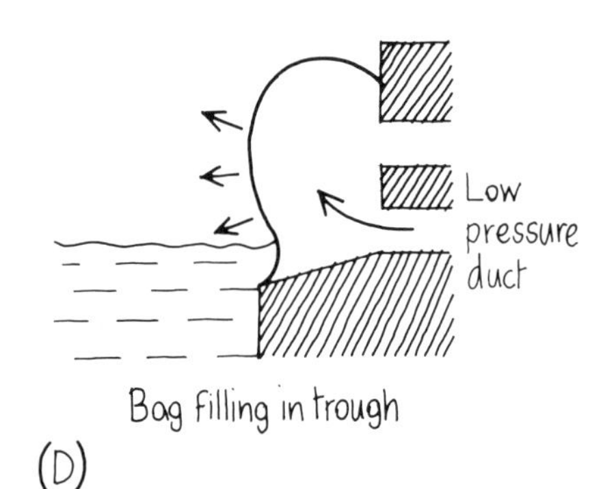

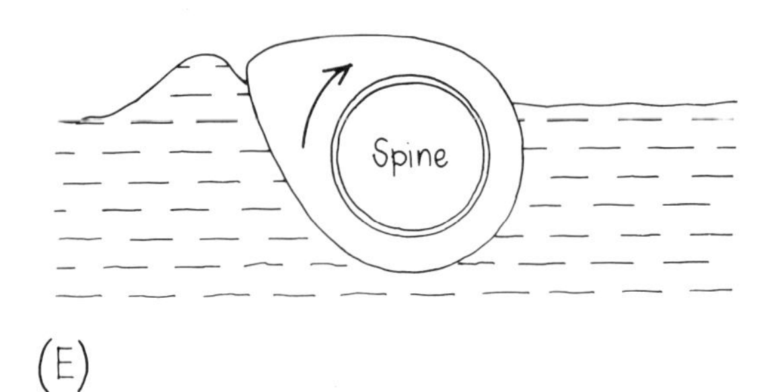

FIGURE 9

SOME ALTERNATIVE CONCEPTS FOR WAVE-POWER DEVICES:

(A) SIMPLE OSCILLATING WATER COLUMN. WATER RISING AND FALLING IN THE SPACE INSIDE THE DEVICE PUMPS AIR THROUGH A WELLS TURBINE WHICH ACCEPTS FLOW IN EITHER DIRECTION

(B) NATIONAL ENGINEERING LABORATORY'S OSCILLATING WATER COLUMN WAVE ENERGY CONVERTOR

(C) BOBBING BUOY. WAVES PASSING OVER THE BUOY INCREASE AND REDUCE THE FORCE IN THE MOORING CABLE THUS CAUSING THE PISTON AND CYLINDER TO PUMP AIR OR WATER THROUGH A TURBINE

(D) LANCASTER FLEXIBLE BAG. WAVE CRESTS SQUASH THE AIR BAG, WHICH ACT LIKE BELLOWS FORCING AIR THROUGH A TURBINE: THE BAGS REFILL IN WAVE TROUGHS

(E) SALTER'S 'DUCK'. CAM-LIKE BODIES OSCILLATE ABOUT A CENTRAL SPINE AS THE WAVES PASS THUS OPERATING HYDRAULIC PUMPS INSIDE THE DEVICE WHICH IN TURN DRIVE TURBINES. (*SOURCE*: FRENCH ,1985; FRENCH, 1988)

Take for example a problem given at the total product level: the conceptual design of devices to harness the energy of the waves. There are very many possible ways in which wave-power might be harnessed, and so this problem offers much scope for inventiveness and creative design ideas – there are over a thousand proposals for utilising wave energy in the patent literature.

One of the simplest wave-power concepts is an oscillating water column which uses the rise and fall of the waves to drive air through a turbine, and many variants of this idea are possible. This and three other design concepts for wave-power devices are shown in Figure 9. Michael French has analysed different possible wave-power devices and has shown that there are twenty-seven basic types. His method for identifying the different concepts will be explained later in Section 7.3.

However, it is unusual for designers, especially in engineering, to be given a problem defined in such general terms. A more typical problem would be to design a component for a particular type of wave-power device to meet a given specification. Figure 10 shows three designs for swivel joints which might be used for connecting undersea cables on a wave-power device. The differences between these designs are in the type of bearing used, and are obvious only to someone with the necessary technical training to interpret the drawings. However, from the viewpoint of the engineers who produced them, designing these joints required a considerable amount of creative work at the conceptual stage involving 'much freehand sketching' before they could be drawn, as shown here, using a computer-aided drafting system (Shahin, 1988).

These are extreme examples. Most conceptual design work is located somewhere between inventing whole new systems and designing minor components. Indeed in this Block we shall be concerned mainly with creative thinking applied to the design of new or improved versions of relatively simple whole products, such as bicycles or vacuum cleaners, or major sub-systems or components of such products. The point of giving these extreme examples is to show that there are alternative solutions to a design problem at every level from the most general to the

FIGURE 10
ALTERNATIVE DESIGNS FOR SWIVEL JOINTS

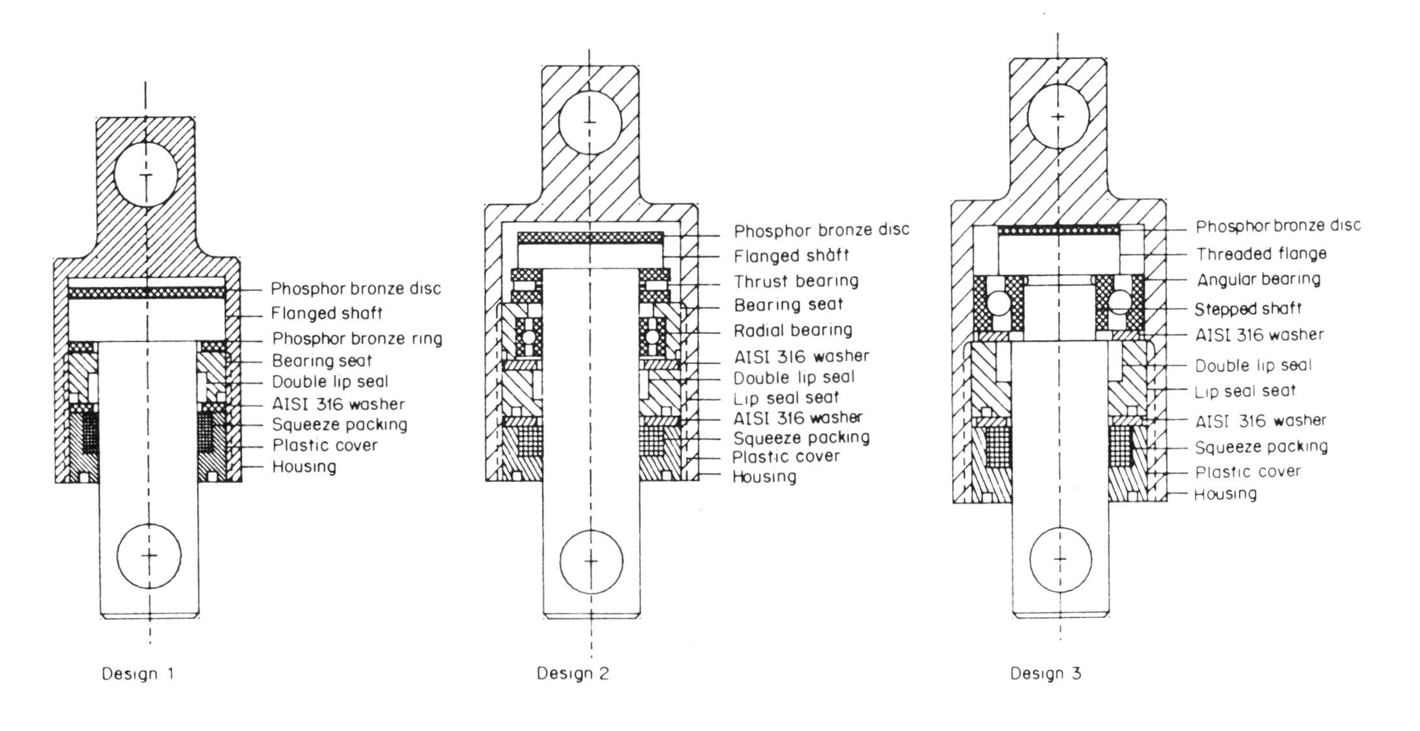

most detailed and that the creative thinking needed to produce worthwhile design concepts is needed at all stages and all levels of design.

Of course, as you should remember from Block 1, design need not proceed in the way I have described. The creation of new ideas by designers sometimes precedes any practical need or commercial demand. Indeed concepts may stem from virtually any stimulus – the personal experience of the designer, a new material, the deficiencies of an existing product, and so on – and not necessarily a brief or specification from a client or manufacturer. The drawback of such approaches is that without a proper specification the designer has no external criteria with which to choose between concepts. Also many designers tend to decide quickly on a particular concept, and then design it in detail without a proper attempt to generate and choose between different options. This too has severe drawbacks, as Stuart Pugh points out:

> This tendency to 'cut and run' is as common with students of design as it is with professionals, and the results can always be bettered. … Remember, you need as many ideas as you can possibly generate – single solutions are usually a disaster.
> […]
> Above all remember that in practice … the wrong choice of concept in a given design situation can rarely, if ever, be recouped by brilliant detail design.
>
> (Pugh, 1991, pp. 69, 73)

EXERCISE ALARM CLOCK CONCEPTS

Suppose you were asked to produce alternative concepts for alarm clocks. Considering different possibilities for the *source of power* used to drive the clock, the way in which the *time is indicated* and the *type of alarm* system, what alternative concepts can you think of? (Note that I am only asking you to generate alternative ideas or concepts for the product as a whole. You do not need to go beyond this into conceptual design proper which would normally require you to indicate how each product concept might work.)

Spend about 10 minutes noting down or sketching your ideas in your Workbook before looking at my suggestions below.

Different possible sources of power include: wound spring, falling weight, electric battery, etc.

Different ways of indicating the time include: hands, electronic display, audio output, etc.

Different types of alarm include: buzzer, bell, various sounds (e.g. music, speech, cock crowing), flashing light, etc.

Combinations of these different options give a large number of possible product concepts from a conventional spring-driven alarm clock with a bell to an electronic clock that speaks the time and tells you when it is time to get up.

This method of producing alternative concepts through combinations of different ways of achieving the basic functions of an object is called **morphological analysis** and is one of the idea generation techniques introduced in Section 7.3 and taught on the audio-cassette for this Block (Audio 2).

1.4 DIFFERENT TYPES OF THINKING INVOLVED IN DESIGN

You have seen that creative thinking is required throughout the design process and that creativity is especially important in the conceptual phase of designing because that is where the basic idea or solution concept is conceived. But, of course, creativity is not the only type of thinking needed in design. In many areas of design, more time and effort may be spent on understanding and analysing the problem in question (perhaps by gathering information or doing calculations) than in thinking up and sketching ideas for solutions. As much effort may be required to decide between possible alternative solutions as was needed to generate them in the first place. This gives us a clue to three essential, but very different, kinds of thinking required in design (and indeed any kind of problem solving). These have been variously described as:

- Analysis, Problem Definition or Problem Exploration;
- Creativity, Solution Synthesis or Idea Generation;
- Decision-making, Solution Evaluation or Idea Selection.

Another way in which to look at these different kinds of thinking is in terms of their use and effect within the design process. Just as creative thinking is usually involved at all stages of design from concept to detail, so are the other types of thinking. In fact the design process may be viewed as successive and repeated cycles through the three types of thinking, as shown in Figure 11.

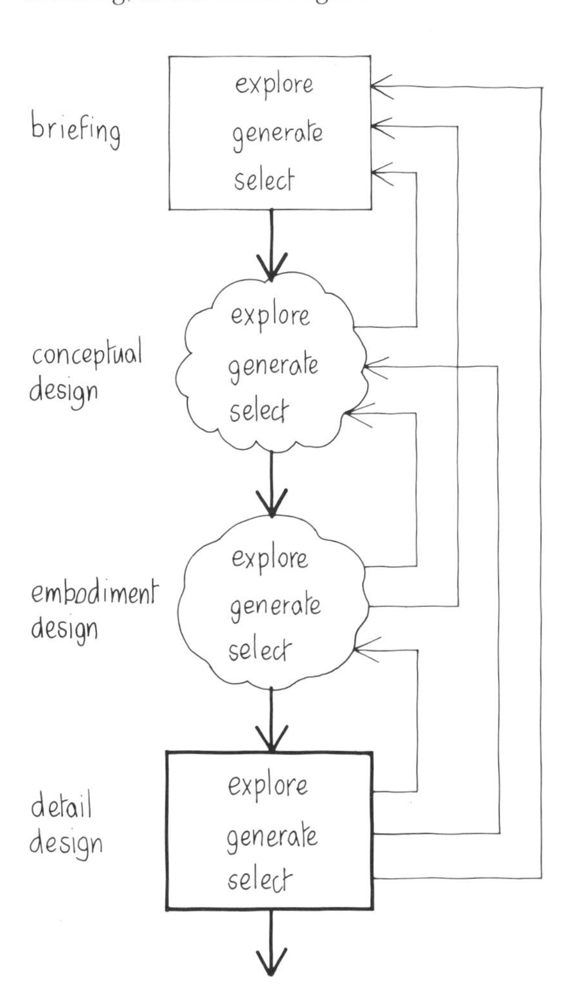

FIGURE 11
THE DESIGN PROCESS SHOWING REPEATED CYCLES OF PROBLEM EXPLORATION, IDEA GENERATION AND SELECTION

Let us look at an example (adapted from Pugh, 1991) of the three kinds of thinking in the conceptual phase of design. The brief is to design a new type of car horn. Before conceptual design is begun, the problem stated in the brief has to be analysed and defined more precisely. This involves the design team in gathering information on existing designs of car horn and other audible warning devices, and identifying different technologies and phenomena for producing sound. This provides the context for generating solutions. The design team also attempts to discover as precisely as possible the performance required for the new design, the range of conditions to which it will be exposed, any constraints for size, weight, cost and so on, to be incorporated into the product design specification.

The next step is the creative part, in which alternative design concepts are generated. Figure 12(A) shows the conventional car horn (number 1, top left) plus thirteen alternative concepts produced by a group of trainee designers. I shall leave until later how designers come up with new ideas and concepts such as these.

Finally it is necessary to evaluate the different concepts in order to choose one (or more) for further development. One way of doing this is to compare each new concept with the existing design using criteria from the product design specification. Figure 12(B) shows an evaluation chart for the various car horn concepts. It reveals concept 5 as having many advantages and few drawbacks compared with the existing design, and so seems like a promising concept for designing a new type of car horn. In this case the evaluation was done by means of group judgement, but as you will see later, choices between concepts often cannot be made without a considerable amount of mathematical analysis and perhaps some testing of models and prototypes.

This example, and the model of the design process in Figure 11, are based on a *systematic* approach to design and do not always correspond to the way in which designers work. As I mentioned earlier, many designers tend to jump quickly from problem to solution without explicitly generating alternative concepts. Other designers start with solutions and use these to explore the problem in hand (the so-called 'solution-focused' approach, which will be discussed in Section 3). But whatever the approach, the overall effect of the process is to narrow down from an ill-defined problem with many possible solutions to a single solution specified in enough detail for manufacture. Within this overall narrowing process there are periods of divergent thinking, in which the problem is explored and solutions are generated, and convergent thinking, during which solutions are analysed in detail and selected.

Creativity is usually considered to involve divergent thinking because it is associated with generating alternative ideas. But creativity can also be convergent if, rather than generating alternatives, it involves synthesis of a wide range of information into a single elegant solution. We will be looking at divergent and convergent thinking again in Section 5. The point for now is not that any particular type of thinking is divergent or convergent, but that all the types are important in design. A good designer (or design team), as you will see later, is an individual or group who can switch between the different modes of thinking; at different times they are able to think divergently and convergently, and are capable of analysing problems, creating solutions and deciding between options.

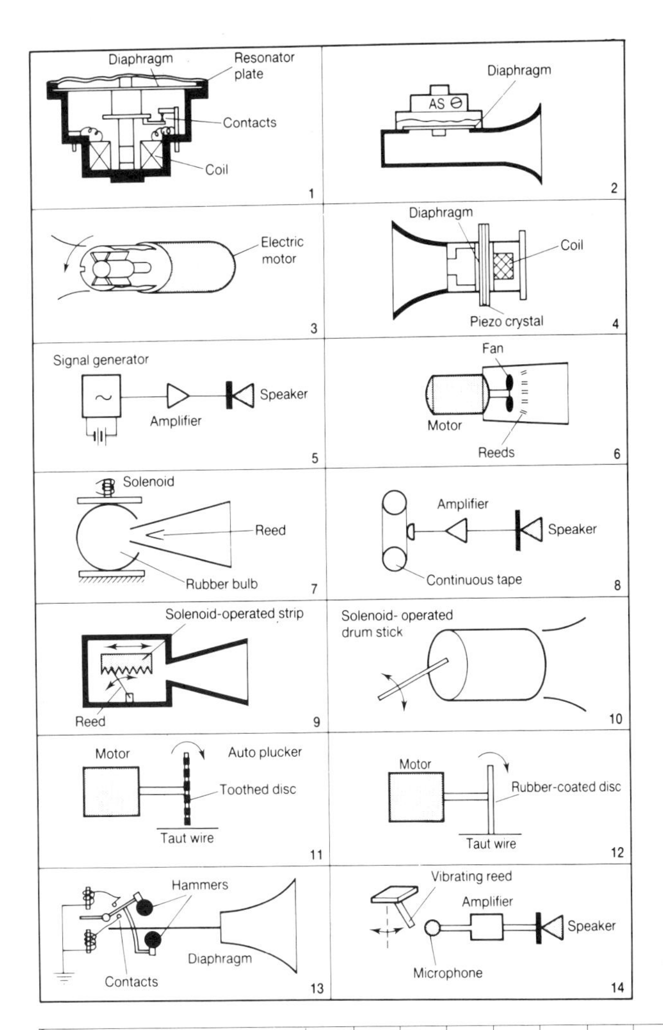

FIGURE 12(A)
ALTERNATIVE COMPARABLE DESIGN CONCEPTS FOR A NEW TYPE OF CAR HORN. THE CONVENTIONAL CAR HORN IS SHOWN TOP LEFT. (*SOURCE:* PUGH, 1991)

Criteria \ Concept	1	2	3	4	5	6	7	8	9	10	11	12	13	14
Ease of achieving 105 - 125 dBA		S	-		+	-	+	+	-	-	-	-	S	+
Ease of achieving 2000 - 5000 Hz		S	S	N	+	S	S	+	S	-	-	-	S	+
Resistance to corrosion, erosion and water		-	-	0	S	-	-	S	-	+	-	-	-	S
Resistance to vibration, shock and acceleration	D	S	-	T	S	-	S	-	-	S	-	-	-	-
Resistance to temperature	A	S	-		S	-	-	-	S	S	-	-	S	S
Response time	T	S	-		+	-	-	-	-	S	-	-	-	-
Complexity: number of stages	U	-	+	E	S	+	+	-	-	-	+	+	-	-
Power consumption	M	-	-	V	+	-	-	+	-	-	-	-	S	+
Ease of maintenance		S	+	A	+	+	+	-	-	S	+	+	S	-
Weight		-	-	L	+	-	-	-	S	-	-	-	-	+
Size		-	-	U	S	-	-	-	-	-	-	-	-	-
Number of parts		S	S	A	+	S	S	-	-	+	-	-	S	-
Life in service		S	-	T	+	-	S	-	-	-	-	-	-	-
Manufacturing cost		-	S	E	-	+	+	-	-	S	-	-	-	-
Ease of installation		S	S	D	S	S	+	-	S	-	-	-	S	-
Shelf life		S	S		S	S	-	-	S	S	S	S	S	S
Σ +	0	0	2		8	3	5	3	0	2	2	2	0	4
Σ -	0	6	9		1	9	7	12	11	8	13	13	8	9
Σ S	16	10	5		7	4	4	1	5	6	1	1	8	3

FIGURE 12(B)
EVALUATION CHART FOR COMPARING NEW CAR HORN CONCEPTS WITH THE EXISTING DESIGN (*SOURCE:* PUGH, 1991)

KEY:
+ MEANS BETTER THAN EXISTING DESIGN ('DATUM')
– MEANS WORSE THAN EXISTING DESIGN
S MEANS SAME AS EXISTING DESIGN OR CANNOT ASSESS

The required mix of skills and abilities of course varies for different types of design problem. Some design problems are sufficiently well defined to be solved by analytical methods alone. Other fairly well understood problems may be solved on the basis of past experience and require only limited creativity. In this Block I am concerned mainly with non-routine design problems whose solution offers the scope for high levels of creative thinking leading to new ideas, concepts and product innovations.

1.5 AUDIO: TYPES OF THINKING IN DESIGN

There is a discussion of the different types of thinking involved in design on Side 1, Part 1 of the audio-cassette for this Block (Audio 2). The discussion is linked to a short creative thinking exercise in the *Audio 2 Study Guide*.

So, before going on to the next section, allow yourself about 30 minutes to listen to Side 1, Part 1 of the audio-cassette and to attempt the exercises in the *Audio 2 Study Guide*.

SAQ 2

What is the role of the product design specification in conceptual design?

SAQ 3

What different types of thinking are involved in design? Why is creative thinking important throughout the design process, but especially at the conceptual stage?

2 INVENTION AND EVOLUTION OF THE BICYCLE

In Section 3 we shall look at some theories of how inventions and creative design ideas arise. But before that I want to give you an overview of the origins and design evolution of the pedal cycle, because it is an example that I shall be using in several places in the Block. The bicycle is of course a rather unusual product in that its basic component parts and design configuration were established towards the end of the last century and, in its diamond-frame, rear chain-driven form, the bicycle has survived virtually unchanged ever since. But because the pedal cycle has become so widespread, familiar and unexceptional, one forgets that its innovation involved many important inventions (from the pneumatic tyre to the roller chain) and literally hundreds of different, and often strange-looking, design configurations. For many years the bicycle was regarded as an outstanding example of a 'type object': the final end product of a long process of craft evolution. But as you will see, cycle design today is by no means static.

The main purpose of this overview is to introduce you to the notion that inventions and new design ideas occur in the minds of creative individuals at particular times depending on the social, technical and economic environment in which they live. It also shows (as you have already seen in the case of the hovercraft) that what happens to those inventions and new ideas usually depends more on that socio-economic and technical environment than on the creativity of the individual. The overview of cycle history also illustrates several other points you have been introduced to in Section 1 regarding the evolution of a novel product: starting with the invention of the initial concept and its main components, through the creation many design variants and the emergence of a 'dominant' design, to incremental improvements to materials, components and methods of manufacture.

2.1 INVENTION OF THE PEDAL CYCLE

The origins of the pedal cycle are disputed. Certainly the possibility of a light road vehicle propelled by human muscle-power had been considered centuries before the practical emergence of the bicycle in the nineteenth century. Indeed, various four-wheeled human-powered vehicles had been built since ancient times.

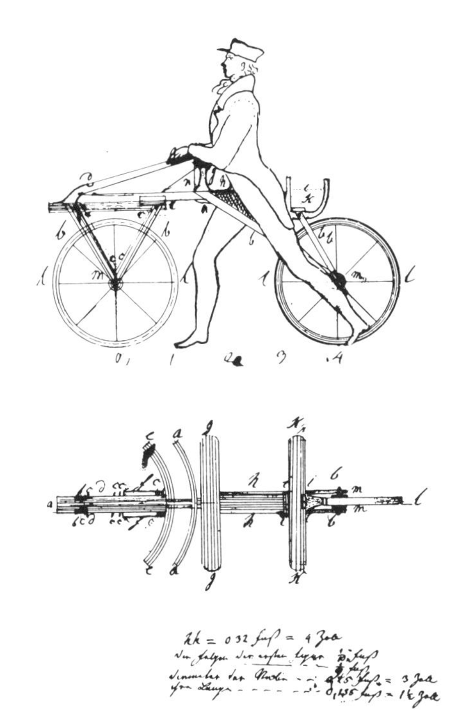

FIGURE 13
THE DRAISIENNE OR STEERABLE ADULT HOBBY HORSE WAS INVENTED IN GERMANY IN 1817 AND PATENTED IN 1818

THE DRAISIENNE

So where did the remarkable idea of a human-propelled, two-wheeled, steerable vehicle on which the rider balanced come from? No one is sure, but what is not disputed is that Karl Freiherr Drais von Sauerbronn, a German baron who worked as a forester, inventor and professor of mechanics, applied for a patent on such a vehicle in 1817 which was granted in 1818 (Rauck, 1983). When the vehicle was demonstrated in France in 1818 it was nicknamed the *Draisienne*. It consisted of a wooden frame to which two iron-tyred, wooden wheels were attached by brackets, the front wheel being steered by means of a handlebar fixed to the top of the front fork. The rider propelled the vehicle by taking long strides with each foot in turn against the ground (Figure 13). Drais, who studied mathematics, physics and architecture at the University of Heidelberg, had been interested in improving the efficiency of human movement and transporting people and goods without horses for several years before he invented the Draisienne. In 1813 he tried to patent a four-wheel crank-driven vehicle propelled by two to four people. When this patent was refused because the vehicle was judged impractical, Drais began to consider alternatives. He probably got the idea for a

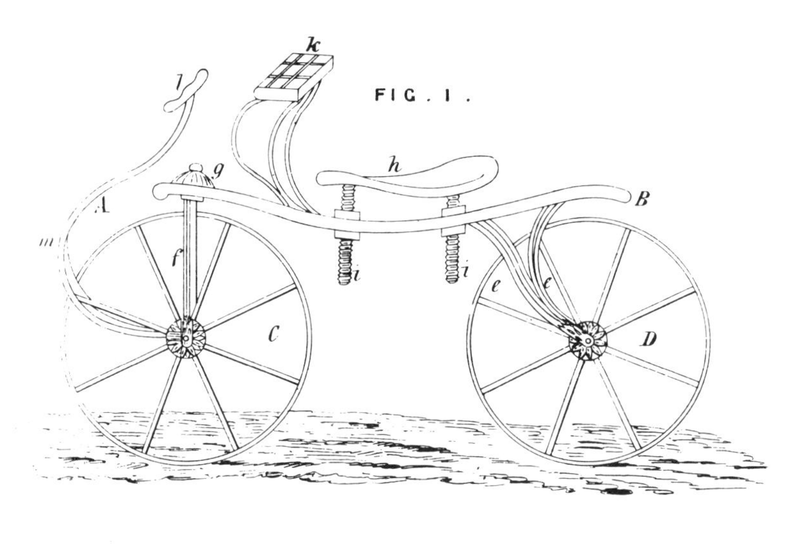

FIGURE 14
AN IMPROVED VERSION OF THE HOBBY HORSE CALLED THE 'PEDESTRIAN CURRICLE' PATENTED BY JOHNSON IN ENGLAND IN 1818

two-wheeled vehicle from the hobby horse, a wheeled toy used by upper class children of which a few had been built for adult use in the eighteenth century. Drais' contribution was to make the adult hobby horse into a practical vehicle by making it lighter and adding steering.

This illustrates one of the most basic sources of creative thinking: *transferring* an idea from one application to another, and, if necessary, *adapting* it. But it is important to realise that someone who was not as inventive as Drais is unlikely to have thought of adapting the hobby horse for practical transport.

However, the Draisienne was only used in a limited way for transport. It mainly caught the imagination of the rich as something to ride for entertainment. What might have remained as an invention thus became a successful product innovation and various designs were made and sold in France, Germany, America and England (Figure 14). The craze however soon faded, and the adult hobby horse had virtually disappeared within a few years of its introduction.

THE MACMILLAN BICYCLE

The first real bicycle, propelled without the rider touching the ground, was conceived and built by a young Scottish blacksmith, Kirkpatrick Macmillan, around 1840. Macmillan made a copy of an adult hobby horse from one he had seen locally and used it to ride in the area of Courthill smithy where he lived. Eventually he had the idea of fitting a system of foot-driven levers linked by rods and cranks to drive the rear wheel and built a second machine incorporating that mechanism (Figure 15). He became an expert rider, and in 1842 rode the 130 miles to Glasgow and back on his bicycle.

Because there is no record of how this key invention in cycle history came about, it is only possible to speculate.

How do you think Macmillan might have come up with the idea for his bicycle?

Macmillan was probably interested in vehicles, as indicated by his copying an adult hobby horse for his own use. He would no doubt have discovered that a moving two-wheel vehicle can stay upright without the feet touching the ground. He presumably became dissatisfied with propelling his hobby horse along with his feet and sought ways of improving it. As a blacksmith he would have been familiar with various devices operated by rods and cranks and, given his practical skills, was able to devise a system for driving the hobby horse mechanically.

FIGURE 15
TREADLE-DRIVEN BICYCLE INVENTED AND BUILT AROUND 1840 BY KIRKPATRICK MACMILLAN, A SCOTTISH BLACKSMITH

In this case we see (or at least have guessed at) another basic source of creativity and invention, namely an individual's desire to improve something, coupled with the skills to come up with a practical solution, plus – as before – the transfer of ideas from one application to another.

However, as you have seen, being creative or inventive is not enough to produce a successful innovation. Macmillan did not patent his invention, and, although his ride to Glasgow caused great local excitement, he did not realise the importance or commercial potential of his bicycle. Macmillan made only one machine, and, although he helped a few others make their own versions of his invention which were sold in limited numbers, the first bicycle was never a successful product innovation.

FIGURE 16
ONE OF THE MICHAUX COMPANY'S VELOCIPEDES OF THE 1860S

THE VELOCIPEDE

Credit for the invention of the first 'proper' pedal cycle has been given to two Frenchmen, Pierre and Ernest Michaux. The Michaux's *Velocipede* was invented around 1860 and introduced commercially in France in 1861. It consisted essentially of a lightweight Draisienne without an arm or chest rest, but with pedals and cranks fixed to an enlarged front wheel (Figure 16). Once again there is no definitive record of how the Michauxs invented the pedal-driven bicycle, but it has been said that they got the idea from thinking about the crank and handle of a grindstone. Whether or not this is true, it exemplifies another common stimulus to creative thinking, the use of an *analogy* or *similar situation* to provide an idea.

The pedal-driven velocipede was a significant advance over the old Draisienne and became very popular in France, England and America up to 1870. It was the first type of cycle to be manufactured in large numbers, and its innovation launched the cycle industry in 1863 with its origins in Paris. The velocipede also triggered the process of evolution that led to the familiar diamond-frame 'safety' bicycle, and its various modern derivatives. Let's look next, therefore, at this process of evolution, this time without examining how each step came about.

2.2 NINETEENTH-CENTURY CYCLE EVOLUTION

A conventional bicycle is an assembly of a very large number of components, ranging from frame tubes and brake cables to spoke nipples and ball-bearings. Hardly any of these existed at the time of the Michaux's Velocipede. Most, including key cycle components such as wire-spoked wheels, chain drive and pneumatic tyres, were invented and developed during a thirty-year period of innovation that started in France and England around 1865. These various components may of course be arranged in different ways to produce different cycle designs, the most common being the standard diamond-frame configuration.

In parallel with the rapid development of cycle components, an amazing variety of different cycle designs were created and introduced. In his definitive book on cycle design, Archibald Sharp (1896) provides a comprehensive catalogue of cycle types classified according to the number of wheels, the method of steering and the means of transmission. In all, Sharp identified twenty-four distinct types of bicycle, twenty-one types of tricycle, and several types of monocycle, dicycle (wheels side-by-side) and multicycle (four or more wheels) that were invented and manufactured at some time during the period from 1865 to 1890. You can get a feel for some of the strange vehicles that were designed from Figure 17. You can also see a few of these in action in Video 3, Section 1, 'The evolution of the bicycle'.

FIGURE 17

A SMALL SELECTION OF SOME OF THE HUNDREDS OF DIFFERENT CYCLE DESIGNS PRODUCED BETWEEN 1865 AND 1890. THEY REPRESENT MANY OF SHARP'S BASIC CYCLE TYPES

With hindsight it is perhaps difficult to understand why, during that period, so many different cycle design configurations were tried out before the diamond-frame bicycle became dominant. One reason for this great variety seems to be that, given the opportunity in a rapidly changing field and booming market, designers, inventors and manufacturers experimented and allowed their imagination free rein, often without an understanding of the technical or design principles involved. Another reason was that, by making use of standard components, any enterprising individual or small manufacturer with only a limited amount of capital could develop and launch a new design of cycle on to the market. All these different designs, and their components, had then to be subjected to the test of time and practical use.

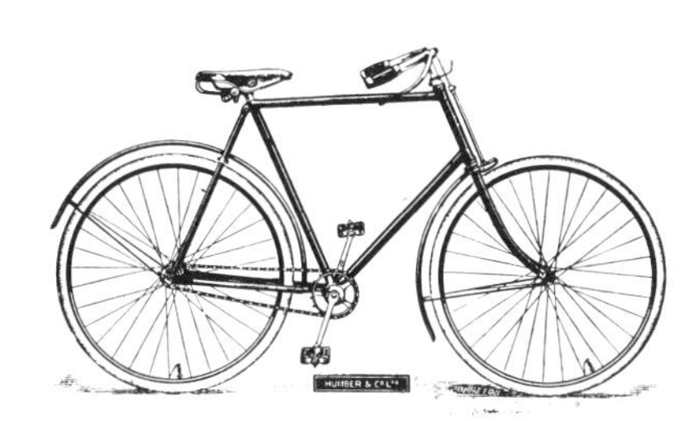

FIGURE 18
HUMBER DIAMOND-FRAME REAR CHAIN-DRIVEN BICYCLE OF THE MID 1890S FITTED WITH PNEUMATIC TYRES. BY THEN THIS DOMINANT DESIGN HAD DISPLACED VIRTUALLY ALL OTHERS. VERY SIMILAR MODELS ARE STILL IN PRODUCTION TODAY

In retrospect, the diamond-frame design may appear simple and obvious (Figure 18); but its apparent simplicity is deceptive. As you will see in Section 4, the bicycle and its components embody many quite complex and subtle ideas in structures, mechanics and ergonomics. Nonetheless, Sharp, writing a decade after the first appearance of the diamond-frame safety bicycle, is critical of the often untutored enthusiasm of Victorian cycle inventors, designers and manufacturers. He comments:

> ... till a few years ago, a great variety of bicycles were on the market, many of them utterly wanting in scientific design. Out of these, the present-day rear-driving bicycle, with diamond frame [...] – the fittest – has survived. A better technical education on the part of bicycle manufacturers and their customers might have saved them a great amount of trouble and expense.
>
> (Sharp, 1896, p. vi)

This raises a question which we will consider later in Section 5: the extent to which theoretical understanding and technical knowledge, as well as creativity and craft skill, are required to produce a worthwhile new design.

We see, therefore, that cycle evolution in the nineteenth century proceeded by a process of inventive thinking, trial and error experimentation and survival of the fittest (or perhaps more accurately non-survival of the less fit) until by the mid-1890s almost all the early types had been displaced in favour of the dominant diamond-frame safety bicycle.

2.3 TWENTIETH-CENTURY CYCLE EVOLUTION

Does this mean that cycle evolution stopped at the end of last century? Certainly not! Human-powered transport has continued to fascinate inventors and designers and, as a result, many ideas for novel cycle designs and for new cycle components and accessories have been developed (or redeveloped) in this century.

Recurrent twentieth-century ideas in cycle design are vehicles in which the rider sits in a 'recumbent' position in order to reduce air resistance and improve comfort (Figure 19); cycles made of lightweight and durable materials such as aluminium alloys (Figure 20) or plastics; and bicycles which fold up for portability and ease of storage (Figure 21 shows one example and on Video 3, Section 3, 'The challenge of the portable bike' you will see several other designs).

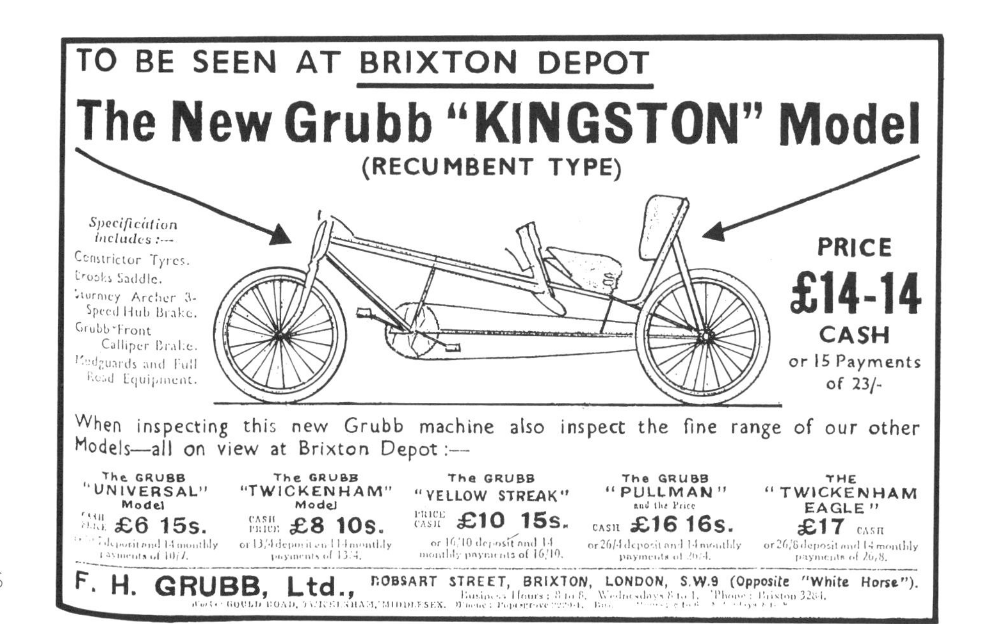

FIGURE 19
ADVERTISEMENT FOR A GRUBB RECUMBENT BICYCLE OF THE 1930S. THE DESIGN ALLOWS THE CYCLIST TO PUSH AGAINST THE SEAT BACK WHILE PEDALLING, AND WIND RESISTANCE IS LOWER

FIGURE 20
BICYCLE WITH A PRESSED ALLOY FRAME DESIGNED BY B.G. BOWDEN AND SHOWN AT THE 'BRITAIN CAN MAKE IT' EXHIBITION IN 1946

FIGURE 21
WINNER OF A 1979 'BIKE OF THE FUTURE' COMPETITION WAS THIS PROTOTYPE FOLDING BICYCLE WITH SHAFT DRIVE, PLASTICS WHEELS AND AN ALUMINIUM OR MAGNESIUM ALLOY FRAME

However, the classic diamond-frame bicycle has proved itself over the years to be so well adapted to its function and use that very few novel types of cycle invented in this century have managed to become successful innovations. This is partly because many of the new designs did not offer enough advantages over the classic bicycle, but also because (as you will see on the video) innovation has been resisted by the cycling world. What has happened instead is that the bicycle has undergone (and continues to undergo) a process of gradual evolutionary development. In the continuing search for strength combined with lightness and efficiency, the turn-of-the-century 'gentleman's bicycle' has evolved into the modern lightweight racing or touring machine. This has involved the development and adoption of improved materials, especially alloy steels, which have allowed stronger, lighter frame tubes, plus the invention and refinement of components and accessories such as gears, lights, brakes and pedals, many of which are now made of aluminium or even titanium. The untrained eye may not even be able to discern some of the more subtle developments, for example, the use of frame tubes on high-quality lightweight bicycles that are 'butted', i.e. thinner-walled in the middle than at their ends. Nevertheless the overall result is a very different product: contrast the nineteenth-century roadster of Figure 18 with the modern lightweight of Figure 22; not only are the materials and components different, but the angles of the diamond frame are steeper in the modern racing machine.

FIGURE 22
MODERN LIGHTWEIGHT BICYCLE WITH BUTTED ALLOY STEEL FRAME AND DERAILLEUR GEARS SUITABLE FOR RACING

Top-class sports bicycles are not designed for the mass market, and bicycles with ordinary carbon steel tubes continue to be manufactured in very large numbers all over the world. The main difference is that new accessories and components, such as dynamo lighting, cable-operated brakes and plastic saddles are now available, and even the most basic model of bicycle may now be decorated with colourful paintwork and transfers.

There are, however, some notable exceptions to the twentieth-century pattern of gradual evolutionary developments in cycle design. In particular, the invention in 1959 by Alex Moulton of a small-wheel bicycle with rubber suspension and a cross-frame configuration stimulated a number of design innovations from the 1960s onwards. The thinking behind the Moulton bicycle and its small-wheel successors is discussed in Section 6.1 and shown on Video 3, Section 1, 'The evolution of the bicycle'.

Also discussed in Section 6 and on the video is the emergence of the mountain bicycle with its heavy-duty frame and brakes and wide tyres. This type was originally created for off-road riding and racing down mountain tracks, but since the mid 1980s the mountain bike has dominated commercial cycle design and component innovation.

Competitive sport has always provided a stimulus for cycle inventors and designers. Recently this has led to many innovations, including 'low profile' racing bicycles with a sloping frame and carbon fibre disc wheels (Figure 23) and streamlined 'human-powered vehicles' (HPVs) with two, three or more wheels (Figure 24). The low profile bikes were designed for the 1984 Olympics and, as you will see on the video, HPVs are at present little more than experimental prototypes designed to break speed records. But usually the process of innovation, if it is to take a radical new

FIGURE 23
LOW-PROFILE BICYCLE, WITH CARBON FIBRE DISC WHEELS AND SLOPING FRAME TO REDUCE AERODYNAMIC DRAG, DESIGNED FOR OLYMPIC RACING AND BREAKING SPEED RECORDS

direction, starts, as in the last century, with an upsurge of experimentation with unusual and improbable designs. By the early 1990s there were already several commercially available recumbent bicycles and tricycles designed for road use, including HPVs with complete bodyshells: Figure 25 shows one example.

Electrically-assisted cycles provide an example of an innovation whose development in Britain was for many years constrained by government regulation. In 1983 new Road Traffic Regulations were introduced that permitted electrically-assisted cycles, with a maximum speed of 15 mph, to be used on the roads by anyone at least fourteen years old. Following the new regulations several models of electrically-assisted cycle appeared on the market. Probably the best known of these was the Sinclair C5. However, like the C5, none of these designs has so far proved to be a commercial success. The reasons for the failure of the C5 were discussed briefly in Block 2, and will be analysed in detail in Section 8 of this Block.

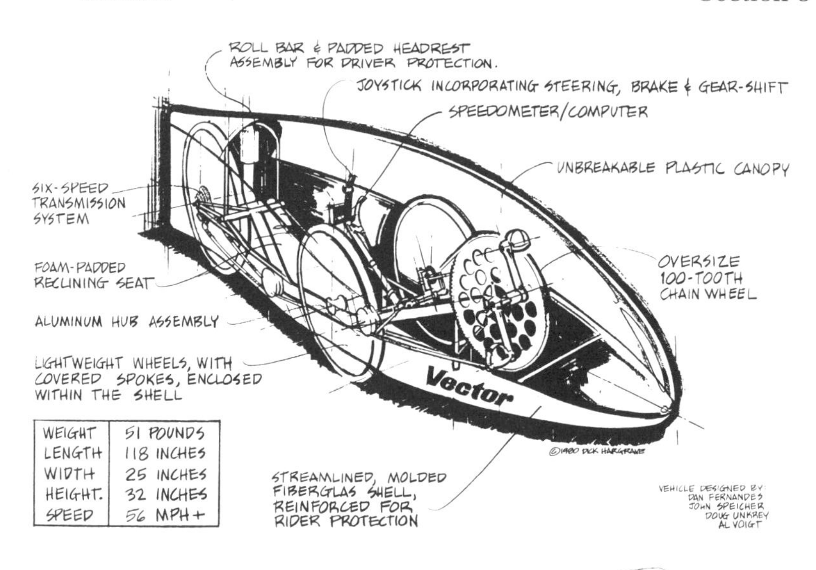

FIGURE 24
VECTOR THREE-WHEELED HUMAN-POWERED VEHICLE WITH STREAMLINED FIBRE-GLASS AND PLASTICS SHELL DESIGNED TO BREAK SPEED RECORDS. FOR MANY YEARS IT HELD THE WORLD SPEED RECORD SET IN 1980 OF OVER 58 MPH. BY THE LATE 1980S NEW DESIGNS OF HPV HAD REACHED SPEEDS OF OVER 65 MPH

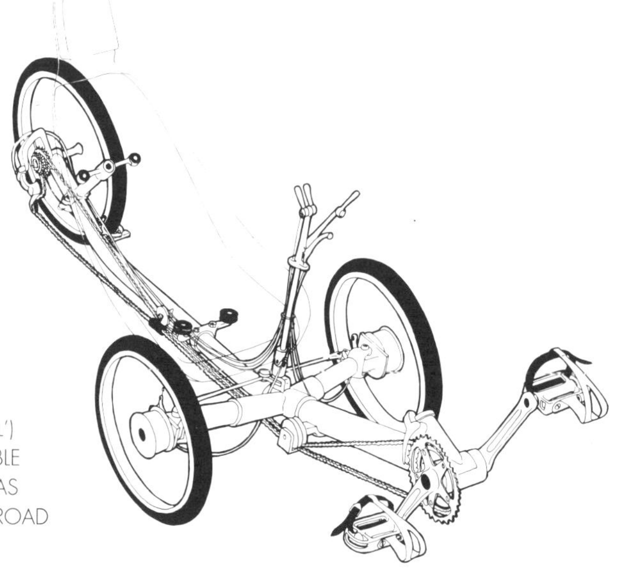

FIGURE 25
BRITISH-DESIGNED WINDCHEETAH SL ('STREET LEGAL') RECUMBENT TRICYCLE FOR ROAD USE AND AVAILABLE WITH AN OPTIONAL GLASS FIBRE BODY SHELL. IT HAS WON MANY PRACTICAL VEHICLE CONTESTS AND ROAD RACES

2.4 VIDEO: THE EVOLUTION OF THE BICYCLE

This brief review has shown that, after the evolution from many different configurations into one 'dominant' design in the nineteenth century and its refinement in the first half of the twentieth century, today we see a renewed diversity in the design of cycles. This pattern of cycle innovation is shown diagramatically in Figure 26 (overleaf).

It is not my intention in the rest of this Block to give you a comprehensive history of the bicycle and its components, or to fill in all the details in Figure 26. Instead I am going to select a few important or interesting designs from that history, mainly to help you understand the thinking processes involved in their creation, but also to show you why they succeeded or failed as innovations.

The examples I shall be using are:

Major nineteenth century innovations – the tangent spoked wheel; the Rover Safety bicycle; the pneumatic tyre; and the Dursley Pedersen bicycle, discussed in Section 4.

Major twentieth century innovations – the Moulton bicycle and its successors; portable bicycles; and the mountain bicycle, discussed in Section 6.

Two contemporary innovations – the Sinclair C5; and the Itera plastics bicycle, discussed in Section 8.

You can see several of these inventions and designs in Video 3, Sections 1 and 3.

You should view the whole of Video 3, Section 1, 'The evolution of the bicycle' now. Then read Section 1 of the *Video 3 Study Guide* and attempt SAQs 4 and 5 below. You will need to look at parts of Section 1 of the video again before studying Sections 4 and 6 of this Block.

SAQ 4

What was the most probable source of the ideas which led to the conception of the Draisienne, the Macmillan bicycle and the Velocipede?

SAQ 5

What is meant by a 'dominant design'? Why did it take many years for the classic diamond-frame bicycle design to become dominant?

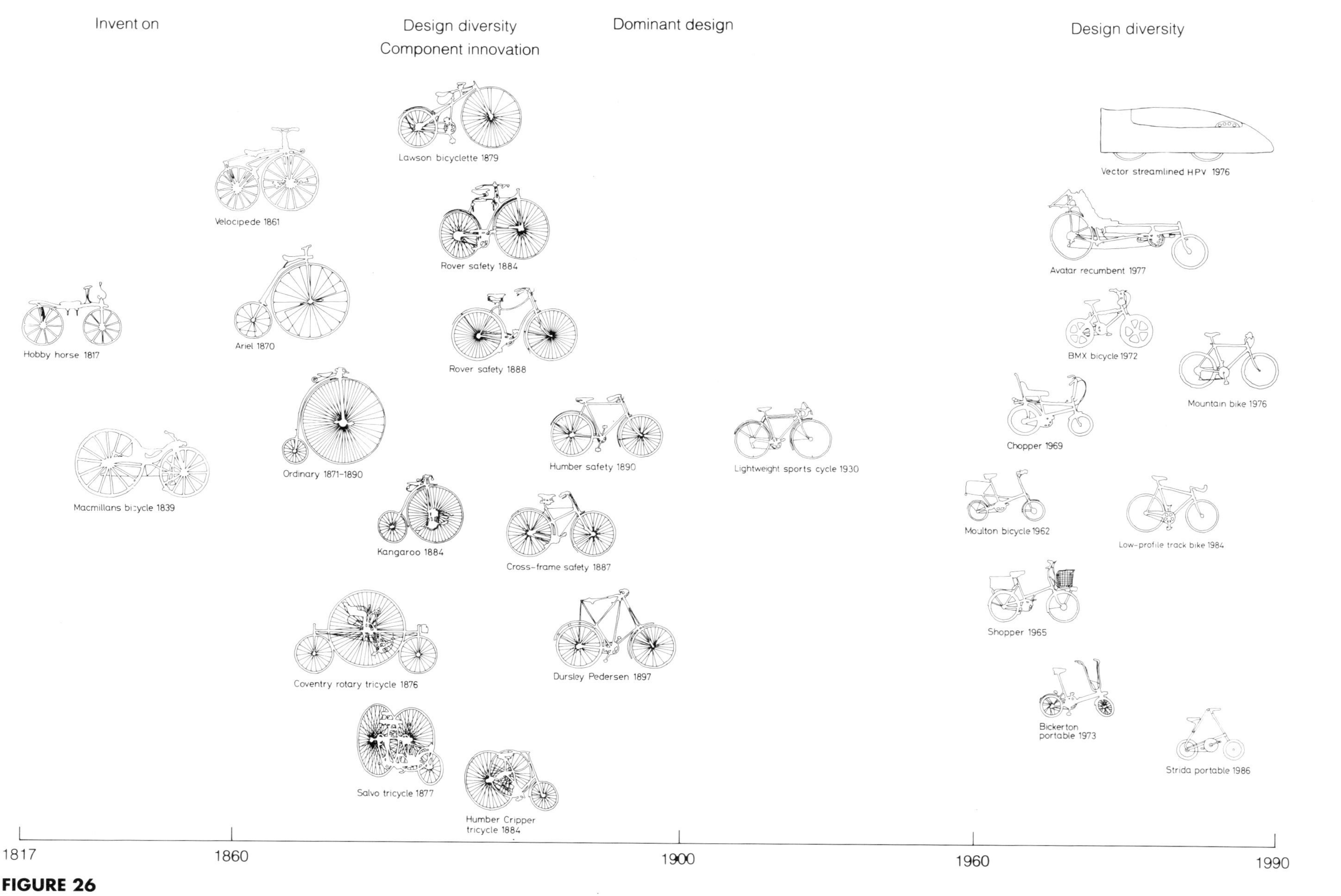

FIGURE 26

EVOLUTION OF THE BICYCLE: FROM THE INVENTION OF THE HOBBY HORSE AND THE DIVERSE DESIGNS OF THE NINETEENTH CENTURY TO THE DOMINANT DIAMOND-FRAME SAFETY BICYCLE AND THE NEW DIVERSITY OF THE LATE TWENTIETH CENTURY

3 CREATIVITY IN THEORY

You have seen that the creation of imaginative and useful new ideas is an essential part of invention and design and that such ideas can arise in a variety of ways. In this section I want to look in a bit more detail at the process of creative thinking and at the different ways in which creative ideas come about. Video 3, Section 2, 'Creativity and innovation' is relevant here, and you will be advised to view it at the end of this section.

Many theories have been proposed that try to explain creativity. Since ancient times creativity has been viewed as a divine gift given to particular individuals. In the nineteenth century this explanation was replaced by the view that creativity was biologically determined by inheritance. In the twentieth century a variety of theories have been put forward which attempt to explain both an individual's ability and motivation to produce creative ideas and the workings of the mind. These range from theories of individual genius to social and psychological explanations of creative thinking. It is not my intention to examine these various theories as no single one provides a satisfactory general explanation. From the practical viewpoint of improving your ability to think creatively in design, the more useful theories are concerned with the process by which problems are solved and new ideas arise, and so it is to this that I shall turn next.

3.1 THE CREATIVE PROCESS

Many of those who have investigated the creative process have identified a typical sequence of stages that people go through while solving problems. In 1896 a famous scientist, Hermann von Helmholz, described his experience of creative problem solving in terms of an initial *investigatory* stage, during which he absorbed the relevant facts; a stage of *rest and recovery*, during which he gave no conscious attention to the problem; and finally a stage of *illumination*, when he had a sudden unexpected insight into a possible solution to the problem. In 1913 the French mathematician Henri Poincaré described his own experience of creative problem solving as first *sensing a difficulty*, followed by a long period of *hard work*. Then came a stage of *unconscious work* in which his mind became receptive to unconscious ideas from which a hypothesis or solution might emerge. Finally the new ideas were applied.

One of the early, and still most widely used, models of the creative process is that formulated by Wallas in 1926 based on the personal accounts of Helmholz and Poincaré.

Wallas identified four stages in problem solving:

- preparation;
- incubation;
- illumination;
- verification.

Preparation is the stage in which the facts necessary for solving the problem are gathered and initial attempts may be made to solve it. Preparation may take minutes or years and often includes considerable effort searching for information, experimentation, trying to find solutions and exploring and redefining the problem.

Incubation is when the problem is set aside, deliberately or otherwise, and no longer given conscious attention.

Then may follow **illumination,** the inspiration or 'flash of insight' in which the solution (or at least the idea on which a solution might be based) suddenly occurs in the mind of the creator, often when they are not working on the problem. This is thought to be the result of the relaxed brain repatterning information absorbed during the period of preparation often after receiving a new piece of information that is perceived as relevant.

Finally there is the stage of **verification** which involves checking the validity of the idea or solution, working out the details and generally developing it for presentation. Again this may take days or years of effort without any guarantee of success.

Although verification is usually the longest, most costly and difficult stage, more interest and attention has been focused on the moment of inspiration than on any other stage in the creative process. Stories, such as how Newton is meant to have had his sudden insight into the laws of gravitation by seeing an apple fall from a tree or Archimedes leaping from his bath crying *Eureka!*, highlight the fact that creative ideas sometimes occur when someone is not consciously attempting to solve a problem. But they disguise the equally important fact that to get that creative idea might have involved years of thinking about the problem and trying to solve it, and may still require many years of effort to get it to work in practice.

The famous example of James Watt's invention of the condensing steam engine highlights this point:

> When given a model of the Newcomen engine to repair in 1763, he soon perceived the problem of inefficiency caused by heat loss in the cylinder's wall. For two years he set the stage by tinkering with the cylinder and trying wooden rather than brass cylinders. Then Watt tells us the act of insight which occurred to him ... while strolling on Glasgow Green, for it was then that he hit on the idea of condensing the steam not in the operating cylinder as Newcomen had done, but in a separate condensing chamber.
>
> Although Watt conceived his brilliant idea of the separate condenser in 1765, it was not until 1769 that he obtained his first patent, and it was more than a decade later, in 1776, that the first Watt engine was brought into commercial use. What happened during the eleven years between Watt's act of insight and the first commercial installation of his engine proves the importance – and difficulty – in transforming an idea ... into a practical innovation.
> [...]
> Watt lacked sufficient capital to devote full-time effort to scaling up his model to an efficient and reliable machine, which involved the solution of many additional technical problems. Watt also lacked the requisite managerial and entrepeneurial expertise. [Matthew] Boulton became the driving force making for the successful introduction of the Watt steam engine: he provided the capital and also brought together the market demand with the creative ability of Watt.
>
> (Kelly *et al.*, 1978, p. 79)

The example of the condensing steam engine supports Wallas' model, of the creative process, but highlights two weaknesses. First, it was based on creative thinking in mathematics and science where a problem may be solved by an individual, whereas in invention and design the stages of developing an idea usually go beyond individual creativity into the realms of innovation, usually involving a large team of people. Secondly, the model fails to mention an important preliminary stage in the creative process, namely that of *problem identification*.

Indeed, the ability to identify (or even seek out) gaps in existing knowledge, or problems that need solving, is considered to be a key aspect of creative and inventive thinking. As you will see later, creative designers are usually people who are sensitive to deficiencies in what exists or to opportunities for change. In design you will see that the identification of a problem to be solved and the way it is defined (often in the brief or specification, as was discussed in Block 2) have a crucial effect on the range and nature of the solutions produced.

In his book *How Designers Think*, Brian Lawson (1980) provides a five stage model of the creative process which includes the initial step of identifying a problem (see Figure 27). This model is useful in defining circumstances under which creative ideas may occur, but Lawson warns that reality is usually more complicated than is suggested by a simple linear model and that processes other than sudden inspiration often produce creative results. So let us next consider some other processes that may be at work.

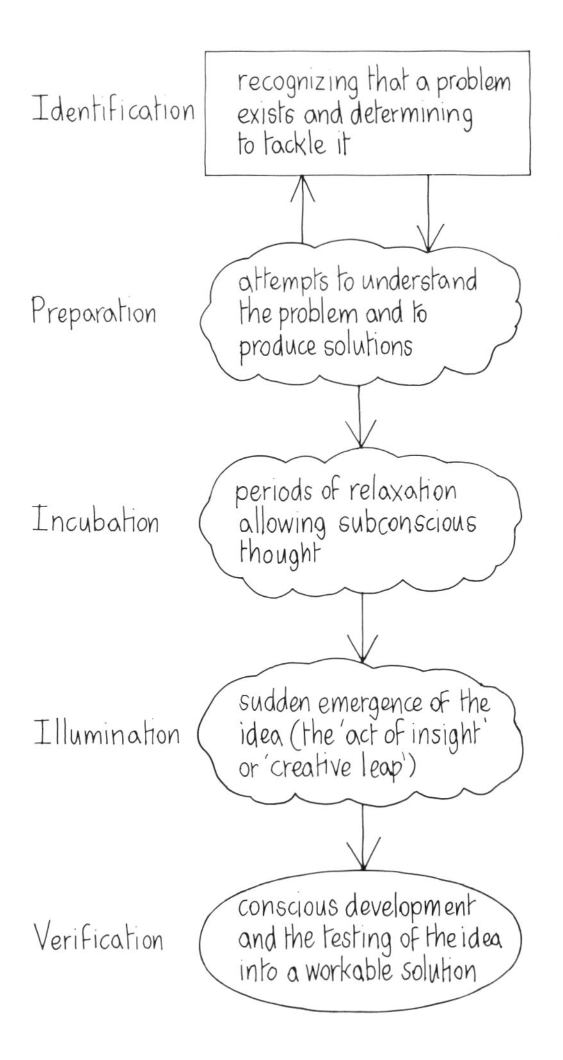

FIGURE 27
FIVE STAGE MODEL OF THE CREATIVE PROCESS. (ADAPTED FROM LAWSON, 1980)

3.2 ASSOCIATIVE THEORIES OF CREATIVITY

Much of the theory concerning the creative process suggests that creative thinking is a matter of forming new **combinations, associations** or **syntheses** of elements of existing knowledge. These combinations and syntheses may arise through a variety of thinking processes: a systematic search for ideas, random and chance events, by analogy, the recognition of relationships between ideas, objects or areas of knowledge that are similar to or different from each other. Indeed the basis of many of the creative thinking techniques introduced in Section 7 of this Block is the deliberate stimulation of associations, or the production of combinations of ideas and elements.

Can you recall any of these processes at work from the examples given in Section 2?

The idea for the pedal-driven bicycle supposedly coming from an analogy or association with a different object (the handle of a grinding wheel).

In his book *The Act of Creation*, Arthur Koestler (1964) argues that the creative act (whether in art, science, technology or humour) involves the synthesis of ideas from two normally unassociated contexts or planes of thought.

The first plane represents all the known and familiar ideas within a particular field. As the mind scans that plane for solutions there are no surprises. The second plane represents another area of knowledge and ideas which, when connected to the first, provides a new solution. Koestler calls the creative linking of ideas from the two planes of thought 'bisociation'.

Koestler cites Gutenberg's invention of the printing press as an example of bisociation. In the fifteenth century, printing was achieved by rubbing paper placed over inked wooden blocks engraved with text or pictures. Gutenberg's attempts to devise a method of printing whole books first required a method of making blocks of type by a means other than engraving. He did this by realising that coins and seals were made by casting and that seals were used for printing. Then came a long period of unsuccessful attempts to improve on the traditional method of printing by placing paper on the type and rubbing. Finally the idea for the printing press came to Gutenberg while he was taking part in the wine harvest. He saw that the pressure that was applied by the wine press could be used to transfer an image from a cast seal or block of type on to paper.

Gutenberg's work of course did not stop at that moment of bisociation. It was the beginning of a long period of intense effort to improve the ink and to develop movable type for the press.

According to Koestler the key creative act was the bisociation of the wine press and the seal to produce the printing press. However, as a general explanation of creativity bisociation is probably too simple because new ideas may arise from combining *more than two* areas of knowledge. Koestler also shows that bisociation is only part of the creative process.

What else was involved in the invention of the printing press?

Gutenberg used analogy (how printing with seals was like printing with engraved blocks). You should also be able to see how his thinking broadly followed the creative process described earlier. (A long period of preparation and incubation followed by sudden illumination and lengthy verification.)

What this means is that the theories of the creative process I have described so far are complementary rather than contradictory. Combination, association and analogy are important generators of new ideas, but as part of an overall creative process. However, I have not yet exhausted the main sources of creative ideas, and in the next section I will briefly consider some more.

3.3 WHERE DO INVENTIONS AND DESIGN IDEAS COME FROM?

If we studied a large number of inventions and new design ideas and considered how they were created we would find a relatively small number of mental processes at work. I have already identified several of the most important ones mainly with reference to historical examples of major inventions. Let's see now how these and other sources of ideas arise in more recent examples of invention and design.

REDEFINING THE PROBLEM

In Block 2 you learned that the way a design problem is defined (e.g. in the brief) will constrain the ideas that might be produced to solve it. It is therefore most important that problems are not defined too narrowly, for example in such a way that they prevent the creation of solutions based on new design concepts or inventive principles. The creative designer will therefore often redefine a given problem or brief in terms of *functions* rather than in terms of a solution.

A nice example of how redefining a simple design problem in functional terms was necessary before a creative solution could be found is provided by David Pye in the following extract from *The Nature and Aesthetics of Design*:

> The author once set about designing a draining rack … for the plates, pot-lids and so forth used by his family while living in a tent. It had therefore to be very small and light. Because he started by thinking "I must design a draining rack" instead of considering what kind of result was wanted, his train of thought was conditioned unprofitably....Any instance of a rack which will support plates must have dimensions conformable with those of the plates and there is a limit below which its size and therefore its weight cannot be reduced.
>
> After prolonged thought the designer realised his mistake and started to consider what result he wanted, namely, a row of plastic plates edge-on in mid air. He then started to search his memory for results of the same class but not necessarily involving similar objects … Doing this is not as easy as it sounds. Because it was not easy his mind ran to a result … suggested by a very obvious association, namely a row of cups hanging on hooks … Thus the thought of plates unearthed the memory of cups.

It was then easy to arrive at the required invention, a thin stick carrying a row of thin wires like cup-hooks; for the desired result was now well in mind, and the objects in it too, the ... plastic plates, which being rather soft at once suggested that holes might be cut in their rims.

(Pye, 1978, pp. 58–9)

What else, apart from the need to redefine the problem in terms of the 'desired result', did you notice about the thought process described by Pye when creating his new plate rack?

To achieve his desired result Pye had to make the mental shift from the principle and concept of supporting the plates from below to suspending them from above. To make this shift he used association to produce a new idea. Notice too that Pye says that searching one's memory to find associations between different objects is not easy; he only managed a fairly straightforward association between cups and plates. Later in Section 7 you will find several techniques which help the mind to find associations between very different objects.

Pye argues that the common practice of defining a design problem in terms of its solution (design a plate-rack) rather than the required functions (drying and storing plates) is often a bad habit because it can hamper creativity. But he also argues that if there are perfectly satisfactory solutions to an established design problem, such as designing a chair, it is often common sense to stick with those rather than to start from first principles. Nevertheless, as you saw in Block 1, and will see in this Block, there is scope for designers to produce innovative solutions to problems even as old as designing chairs and bicycles.

THE PRIMARY GENERATOR

Because there are usually a very large number of possible solutions to a design problem, in order to make the problem manageable, designers often envisage an outline solution before the problem is fully defined or explored. This solution is then tested against the realities of the problem and modified as required, rather than arising (however creatively) out of an exploration of the problem. This tendency of designers to start with solutions – or ideas for solutions – when tackling a problem has been called a **solution-focused** strategy. This is in contrast to the **problem-focused** strategy typically adopted by scientists and engineers, who tend to analyse a problem before coming up with a solution (Lawson, 1980).

What is important here is that the initial solutions that designers create often have their origins in an idea, image or objective arising from the designer's experiences and preferences. The idea or objective that produces this initial solution has been called the **primary generator** (Darke, 1979). The notion of the primary generator arose from studies of how architects tackle design problems. These showed that, when faced with complex design problems and numerous external constraints, architects tended to latch on to a relatively simple idea, image or objective early on to enable them to generate an initial solution. This 'primary generator' might be concerned with practical functions such as wanting to make best use of available light or open space, or more psychological objectives such as wanting to express a particular atmosphere or feeling through a building, or some combination of the two. (You should recall the distinction between the practical and psychological aspects of a design from Block 1.) Often the primary generator is a favourite design idea from another context. For example, the architect Richard MacCormac said about his scheme for a residential building for an Oxford college that his wish to include a raised viewing turret or belvedere was 'the crucial idea that stabilised the whole concept'. The belvedere idea came from his visual memories of a variety of other buildings and other structures which he liked. Indeed, as you will see later in Section 5, probably the most common source of creative design ideas is the repertoire of material that a designer has accumulated from his or her past experience.

In other areas too designers often have primary generators for their solution concepts. For instance, the external form of some of Olivetti's office calculators of the 1970s were generated by their designers' idea of making them psychologically less machine-like by basing the design on the curves of the human body. As you study this Block you will come across other examples.

> Look back at the quotation by Mark Sanders given in Section 1.1 about how he conceived his folding bicycle. What was the primary generator for his solution concept?

> The basic concept was generated by the objective that the folded form should be rather like a 'long thin stick with wheels at one end'. As you will see later in Video 3, Section 3, 'The challenge of the portable bike', Sanders's generating idea was partly practical to do with enabling the bicycle to be wheeled along when folded and partly psychological concerned with the visual image of the folded form.

ASSOCIATIVE THINKING

I have already identified associative thinking as the source of many important new ideas and inventions. Let us look in a bit more detail at some different types. But remember that the way each works is essentially the same – a solution is provided, or an idea for a solution is prompted, by the designer making the mental association between a problem in one field and principles, knowledge or ideas from one or more other fields.

FIGURE 28
THIS CLIPLESS RACING BICYCLE PEDAL, PRODUCED BY A LEADING MANUFACTURER OF SKI EQUIPMENT, IS DESIGNED TO LOCK ON TO A MATCHING SHOE IN A SIMILAR MANNER TO THE STEP-IN BINDINGS USED ON SKIS

Adaptation is probably the most common form of associative thinking. In adaptation an existing technology or solution to a problem in one field is used to provide a new idea for a solution in another. You have seen the process at work in the invention of the first bicycle and the printing press. A simple, almost trivial, modern example would be the adaptation of the pistol grip originally designed for guns for various other objects from power drills to cameras. Another example, shown on the video, is a new design of clipless racing cycle pedal and shoe which lock together in a similar way to the step-in bindings used on skis (Figure 28). You should be able to think of other examples: there are literally thousands. It means that one of the most fruitful sources of ideas in any field is a knowledge of, and willingness to look out for, existing solutions, technologies, and designs in a wide variety of other fields both related and unrelated to the problem in hand. On Video 3, Section 3, 'Creativity and innovation' you will see how James Dyson's innovative designs resulted from his ability to use and adapt technologies from different fields to those he was working in.

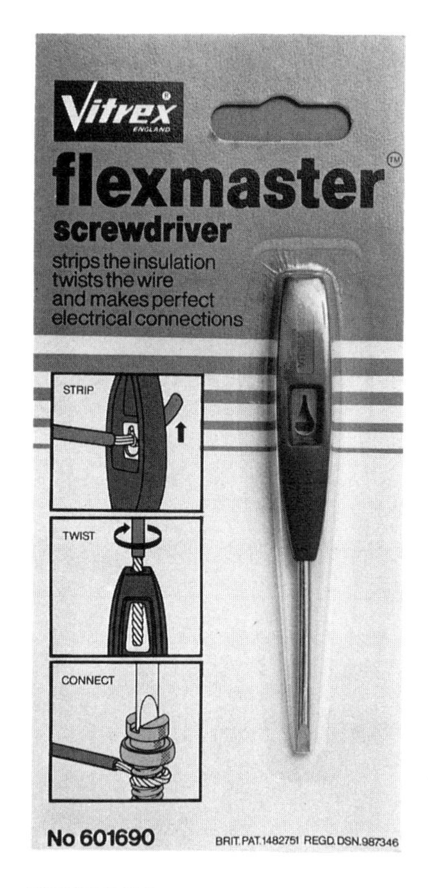

FIGURE 29
BY COMBINING A SCREWDRIVER WITH A WIRE-STRIPPER A USEFUL NEW PRODUCT IS CREATED

Transfer is a form of associative thinking similar to adaptation in which an innovation in technology, manufacturing processes or materials provides the stimulus for a new design or invention. A good example is the transfer of laser technology, originally thought to have few practical uses, to a great variety of applications ranging from navigation and surgery to sound reproduction.

Combination is a particular type of associative thinking that often produces new creative ideas. The most elementary form is where two or more existing devices are combined to produce something new. A simple example might be a combined screwdriver and wire-stripper (Figure 29). A more sophisticated example is a tent designed for use by cyclists. The cycle frame is used to support the front of the tent thus eliminating the need for tent-poles (Figure 30).

Several of the creative problem solving techniques outlined in Section 7 are based on combining existing ideas or objects to produce something new. However, these techniques differ from the 'normal' process of creative thinking in that they deliberately make explicit the process of generating combinations and separate it from the process of choice. It has been said that the *choice* of useful combinations from the mass of possibilities is perhaps even more important in creative thinking than the generation of the combinations:

> ... invention ... takes place by combining ideas.
> [...]
> ... this building up of numerous combinations, is only the beginning of creation ... to create consists precisely of not making useless combinations and in examining only those which are useful and which are only a small minority. Invention is discernment, choice.
> (Hadamard, 1945, p. 30)

I shall return to the importance of choice in creative thinking in Section 5.

FIGURE 30
CYCLE TENT. THE COMBINATION OF BICYCLE AND TENT ELIMINATES THE NEED FOR TENT-POLES THUS REDUCING WEIGHT AND BULK. THIS INGENIOUS DESIGN WON A NATWEST/BP AWARD FOR TECHNOLOGY IN 1989

Analogy – drawing on *similar* situations – is yet another type of associative thinking which, as you have seen earlier, often produces new ideas. *Biological analogies* in particular often provide or stimulate ideas for invention and design. For example, an ultrasonic focusing system for instant cameras was based on the method used by bats to navigate in the dark, and an optical disc for information storage derived from research on the working of a moth's eye. The designer Victor Papanek has used various biological analogies, notably aerodynamic plant seeds, as a rich source of design ideas. For example, he has proposed dropping plastic 'seeds' containing fire-extinguishing powders from aircraft to help put out forest fires in inaccessible areas. Other uses suggested for artificial seeds are reforestation and fish restocking (Papanek, 1972).

There are many examples of plants and animals providing useful design principles in engineering and architecture. An example is a novel bridge based on the vertebrae of a dinosaur that its designers created after a visit to the Natural History Museum in London (Figure 31). Biological analogy provided design ideas for another bridge, the Kingsgate footbridge in Durham. When thinking about the main structural members, its designer Ove Arup later said that he 'saw them like a thing growing out, like a rhubard stem' (Walker and Cross, 1983). In early sketches of the design the concave stems of rhubarb were copied (Figure 32). Later, however, Arup made the supports convex because he realised that water might collect where two concave struts joined at the base. In both these cases the biological analogy provided the initial stimulus for the design rather than the final form.

FIGURE 31
BIOLOGICAL ANALOGY. THE CONCEPT FOR THIS NOVEL ASYMMETRICAL BRIDGE WAS BASED ON THE VERTEBRAE OF A DINOSAUR. THE BRIDGE'S MAIN STRENGTH WOULD COME FROM ONE SIDE OF THE GAP IT SPANS AND THE THRUST WOULD BE CARRIED BY THE VERTEBRAE, STRUTS AND CABLES OF THE SPINE. BEFORE GETTING THE IDEA FOR THE BRIDGE ITS DESIGNERS WERE ALREADY ALERT TO THE USE OF BIOLOGICAL FORMS IN DESIGN FROM READING D'ARCY THOMPSON'S CLASSIC BOOK 'ON GROWTH AND FORM'

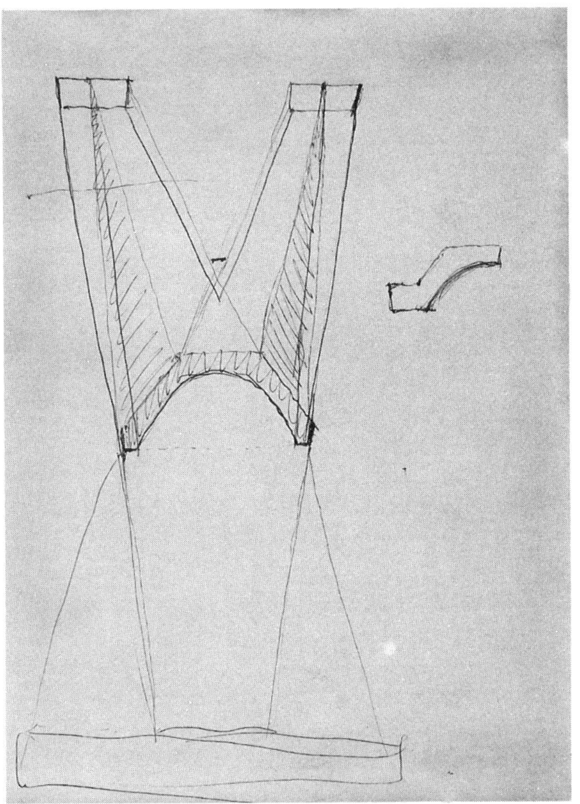

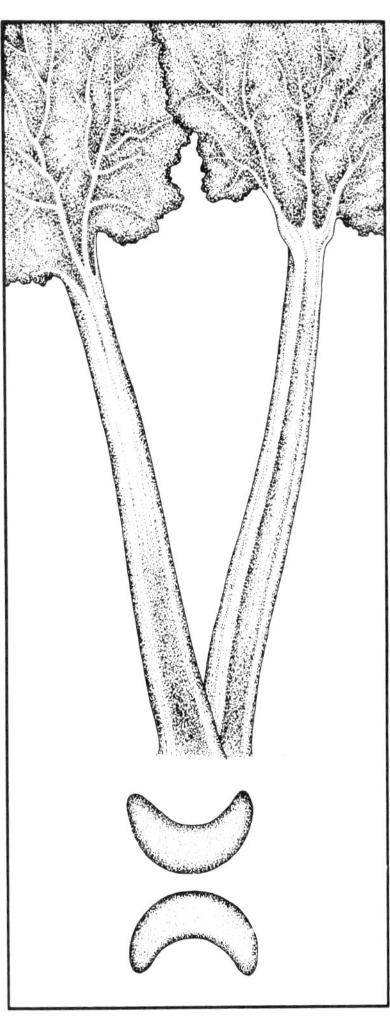

FIGURE 32
BIOLOGICAL ANALOGY. A PAGE FROM OVE ARUP'S SKETCHBOOK SHOWING (IN SECTION AND ROUGH PERSPECTIVE) HIS INITIAL IDEA FOR THE MAIN STRUTS OF THE KINGSGATE FOOTBRIDGE IN DURHAM, CONCEIVED AS AN INVERTED RHUBARB STEM

Chance is another important source of inventions and scientific discoveries. A chance event will sometimes provide the answer to a problem an individual is already trying to solve. In this category is Goodyear's discovery of the process of vulcanising rubber when he accidentally dropped a piece mixed with sulphur on a hot stove. Other chance inventions arise when the individual has no particular problem in mind. An example is 'Velcro' the idea for which occurred when George de Mestral happened to notice how burrs stuck to clothing. (Had de Mestral been actively looking for new ideas for fastenings, Velcro might be seen as the result of biological analogy.)

But, in both types of chance invention, Pasteur's observation that 'chance only favours the prepared mind' is usually true. It is most unlikely, for instance, that Alexander Fleming would have noticed the effect of penicillin mould, which by chance had fallen on to a bacterial culture, had he not previously been experimenting with antibacterial enzymes. A creative individual has therefore to be prepared and ready to make use of chance events to provide solutions to existing problems or to generate new ideas. Indeed many of the creative thinking techniques outlined later in the Block deliberately make use of chance associations to produce new ideas.

EXERCISE ASSOCIATIVE THINKING

Look at the Guided Design Exercise (GDE), which forms part of TMA 03. Spend 10 minutes writing down as many things as you can think of that you associate directly or indirectly with, or have some similarity with, the object that is the topic of the GDE. Don't attempt to evaluate your list or worry if the things you note down (which may include products, plants, animals, etc.) seem odd or even irrelevant at this stage. The list should prove useful later when you are working on the GDE. If your list triggers some ideas for tackling the GDE now, write them down too. Indeed it is useful to keep a section of your Workbook for noting or sketching 'spontaneous ideas' for tackling the GDE whenever they occur.

THEORY AND EXPERIMENT

Associative thinking is often the source of the basic idea or solution concept for an invention or new design which may come in a sudden 'flash of insight'. However, creating this basic idea or concept is often the easiest part; making it work usually requires the hard grind of direct and sustained attempts to tackle a problem.

The development of an idea or concept often involves long periods of experimentation. As you will see on the video, James Dyson developed his Cyclone vacuum cleaner by building and testing literally thousands of models and several prototypes. This experimentation might be based on trial-and-error methods, a theoretical understanding of the problem, or a combination of the two. Edison, for example, experimented with hundreds of different substances in his search for a suitable filament for his electric lamp (Figure 33). Although he used trial-and-error methods Edison's experimentation was by no means unscientific: he systematically sought a filament that would last for hours rather than minutes and which was of high resistance to minimise the current, and hence the cost of transmitting the electricity. Studies of the invention of the telephone have shown that Bell and Edison, who produced designs based on different technical principles, did so by extensive experimentation which derived from their mental concepts about how sound might be converted into fluctuating electric current and their

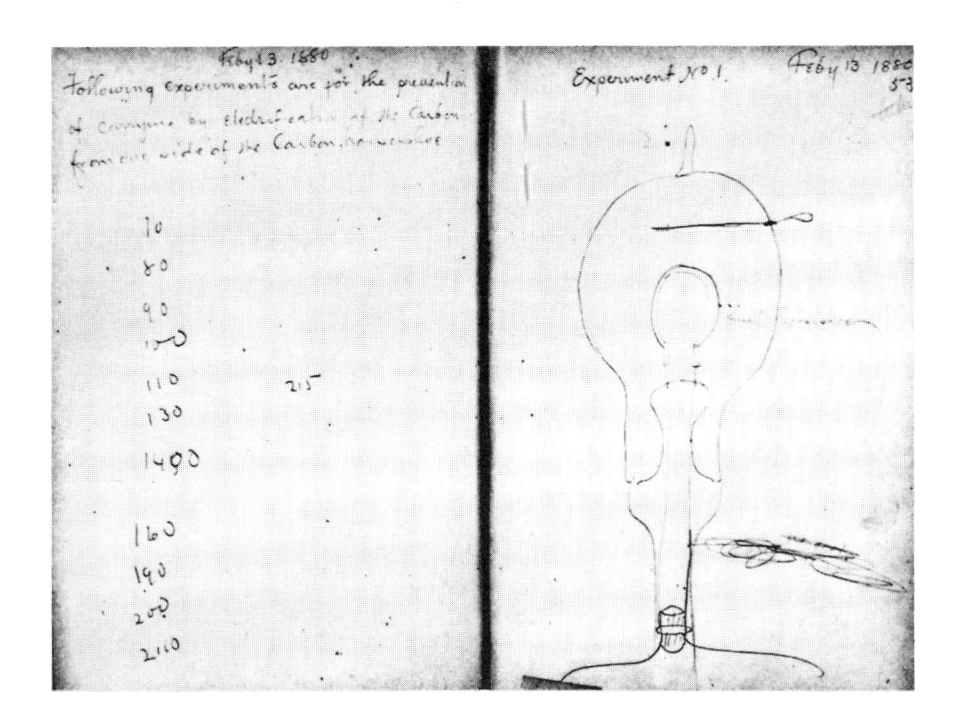

FIGURE 33
A PAGE FROM ONE OF THOMAS EDISON'S NOTEBOOKS SHOWING THE RESULTS OF EXPERIMENTS ON A CARBON FILAMENT ELECTRIC LAMP. EDISON EXPERIMENTED WITH HUNDREDS OF SUBSTANCES IN HIS SEARCH FOR A DURABLE, HIGH-RESISTANCE FILAMENT

practical knowledge of various devices that might perform the desired functions (Gorman and Carlson, 1990). Their inventions were based on both theory and experiment.

Making and experimenting with physical models, prototypes and so on normally is the way that ideas are developed and tested rather than originally conceived. However, some designers like James Dyson also generate ideas through working with physical materials and objects. What starts as germ of an idea is evolved and developed through tinkering with the object itself.

AN EXAMPLE: THE INVENTION OF XEROGRAPHY

The invention of xerography by Chester Carlson is an excellent example of the use of scientific and technical knowledge plus sustained experimental effort to conceive and develop an idea.

In the early 1930s Carlson saw the potential for a new system of copying documents that was quicker and cheaper than photography or hand copying. He decided from the start to look for a way of using light directly instead of the photographic methods being pursued by the large photographic companies. His work as a patent lawyer enabled him to do an extensive search of the patents and other literature for ideas. As Carlson himself later said, 'Things don't come to mind readily all of a sudden, like pulling things out of the air. You have to get your inspiration from somewhere and usually you get it from reading something else.' (Quoted in Mort, 1989.) After identifying some promising ideas he started experimenting and following much effort Carlson, together with an engineer he had hired, produced the first xerographic print on a sulphur-coated plate in 1938. The invention combined two effects: photoconductivity, in which the conductivity of a material is enhanced by exposure to light, and electrostatic charging to attract a powder and produce an image. Carlson filed a patent on what he initially called 'electrophotography' in 1939 which was granted in 1942 (Figure 34).

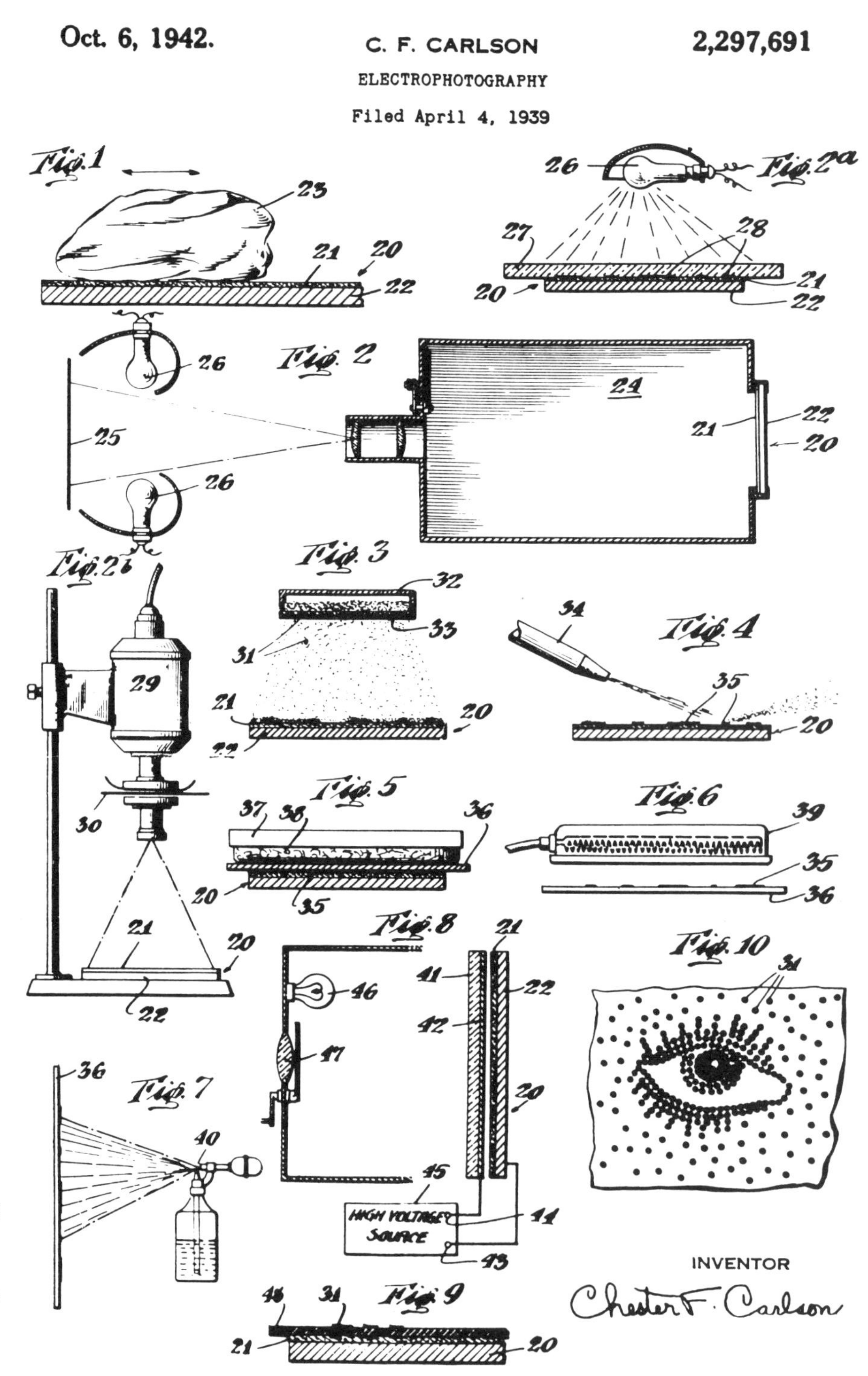

FIGURE 34
FIRST PAGE FROM CHESTER CARLSON'S 1939 PATENT APPLICATION ON 'ELECTROPHOTOGRAPHY' WHICH ESTABLISHED THE ESSENTIAL PRINCIPLES OF XEROGRAPHIC COPYING MACHINES

Several years of frustrating effort followed before Carlson managed to build the first automatic copying machine and obtain backing to develop his invention into a commercial product. It was not until 1949 that the first commercial XeroX machine was introduced and the early 1960s before xerography began to take off commercially. In the twenty years between Carlson's patent and the launch of the hugely successful XeroX 914 copier, much engineering skill and many new developments in materials, chemistry and electronics had been applied to evolve the invention into a commercial innovation.

Although the invention of photocopying demonstrates the importance of sustained effort based on theory and experiment, it also shows that several other sources of creative ideas discussed in this section played a part. To finish this section, see if you can identify these other sources.

The idea of copying using light was the primary generator. The invention itself depended on the combination of photoconductivity and electrostatics. And no doubt chance was involved too.

3.4 VIDEO: CREATIVITY AND INNOVATION

On Video 3, Section 2, James Dyson talks about his approach to invention and innovative design. In particular he discusses how he got the idea for and then designed and developed two innovative products: a new type of wheelbarrow with a ball-shaped wheel (Figure 35) and a novel type of vacuum cleaner which operates using the cyclone principle (Figure 36). The video demonstrates the importance of physical modelling when attempting to get an inventive idea to work effectively. Dyson also discusses the difficulties he faced in getting his inventions manufactured and marketed as commercial innovations.

FIGURE 35
JAMES DYSON WITH THE BALLBARROW. THE BALL-SHAPED WHEEL IMPROVES THE RIDE OVER ROUGH GROUND

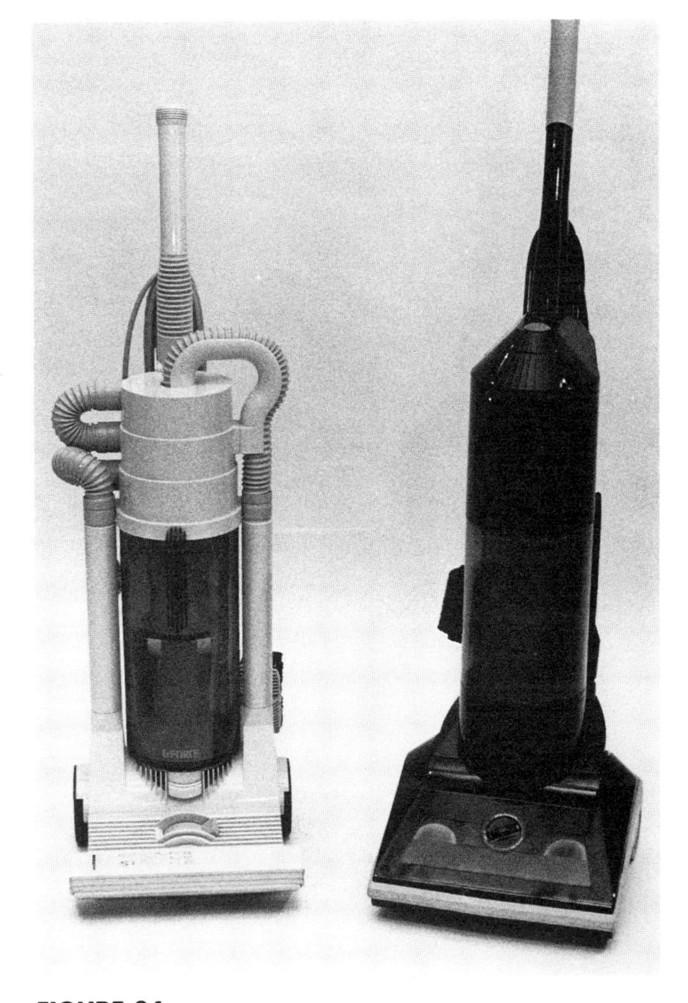

FIGURE 36
TWO VERSIONS OF THE CYCLONE VACUUM CLEANER. THIS INNOVATIVE PRODUCT OPERATES ON A COMPLETELY DIFFERENT PRINCIPLE TO CONVENTIONAL VACUUM CLEANERS

You should read Section 2 of the *Video 3 Study Guide* and view 'Creativity and innovation' now. Attempt the self-assessment questions in the Study Guide. You may need to watch the video section again when studying Section 8 of this Block.

SAQ 6

What are said to be the main stages of the creative process? Does this theory of creativity support the famous saying of the inventive genius Thomas Edison that: 'genius is 1 per cent inspiration and 99 per cent perspiration'?

SAQ 7

Outline different types of associative thinking, giving an example for each. What is common to all these creative thinking processes?

SAQ 8

What is meant by the 'primary generator' in design? Why do designers often start from a primary generator?

4 CREATIVITY IN PRACTICE: NINETEENTH-CENTURY BICYCLE DESIGN

In this section we shall examine four examples of cycle innovations that occurred in the nineteenth century. Three, the Rover Safety bicycle, the tangent-spoked wheel and the pneumatic tyre, were crucial steps in the evolution of the modern diamond-frame bicycle. The fourth, the Dursley Pedersen bicycle, was a novel design that succeeded initially but quickly died out in competition with the dominant diamond-frame machine.

The main purpose of this section is to give you some more examples of how inventions and creative design ideas arise. But here I am going beyond just the thinking involved in the creative act that we focused on in previous sections. I want to show that, even for an apparently simple object like a bicycle, innovations do not happen in a single giant step. They are usually the result of someone bringing together concepts, component innovations, materials and manufacturing techniques that have previously been developed, and doing so under circumstances that allow the invention, or new design, to become a successful innovation.

The first sequence of Video 3, Section 1, 'The evolution of the bicycle' is particularly relevant here. If you have not watched it, you should do so before starting Section 4.1 of this text and attempting the in-text questions it contains.

4.1 THE ROVER SAFETY BICYCLE

The Rover Safety bicycle designed by John Kemp Starley in 1885 (see Figure 37) is generally considered to be the key step in the development of the modern diamond-frame bicycle. In this example I want to examine how the Rover Safety bicycle came about as an innovation, and why it proved to be so successful.

As with most major innovations, the Rover Safety arose from an accumulation of ideas developed earlier. Like most successful designs it put those previous ideas together into a configuration that overcame the drawbacks of other designs in a simple, elegant and efficient way.

FIGURE 37
J.K. STARLEY'S ROVER SAFETY BICYCLE OF 1885. THIS DESIGN IS CONSIDERED TO BE THE FIRST PROTOTYPE OF THE MODERN BICYCLE. NOTE THE DIAMOND-SHAPED FRAME, BUT WITH NO SEAT TUBE IT IS NOT A TRIANGULATED STRUCTURE

FIGURE 38
JOHN PINKERTON WITH A BONESHAKER BICYCLE OF THE LATE 1860S

ORIGINS OF THE ROVER SAFETY

To understand how the Rover Safety emerged, it is necessary to know something about the designs that preceded it and about associated developments in the design of cycle components.

You will recall from Section 2 that the first pedal cycle was the Velocipede invented in France around 1860. By 1868 so-called 'bone-shakers' (Figure 38) similar in design to the Velocipede were being made in England by the Coventry Machinists Company. In Video 3, Section 1, 'The evolution of the bicycle' a cycle historian, John Pinkerton, takes up the story and shows you a few of the cycles and cycle components that were important in the evolution of the Rover Safety, and its modern successors.

Let me start, therefore, considering the main steps in cycle and component evolution that led from the bone-shaker to the Rover Safety by asking you to recall the design reasons for those steps.

> What were the main drawbacks of the bone-shaker for the rider? (Look at Figure 38 to remind yourself of its design.)
>
> ---
>
> The 'bone-shaking' ride, due to its rigid cart-type wheels with iron tyres.
>
> Its restricted speed, due to the limited rate at which the front wheel could be pedalled.
>
> Its limited range and hill climbing ability due to its heavy weight and its lack of gearing.

In the search for increased speed and comfort the attention of cycle designers and manufacturers focused on lighter methods of construction and improved components, notably better *wheels*.

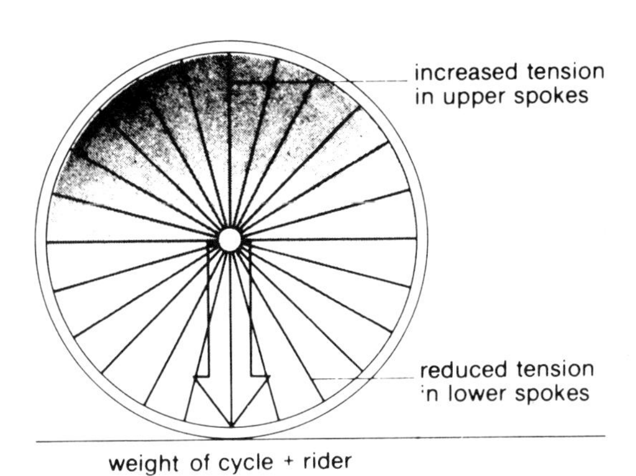

FIGURE 39
RADIAL-SPOKED WHEEL

Cycle wheels

Early cycle wheels were based on carriage wheels with wooden rims, iron tyres and rigid wooden spokes. They were heavy and did not absorb vibrations from the road, hence the term 'bone-shaker' for the cycles equipped with such wheels.

In 1868 the idea of a 'suspension' wheel with flexible wire spokes was patented by E. A. Cowper. In the wire-spoked wheel, the weight of the rider and cycle frame is 'suspended' from the rim through the upper spokes. The wheel remains circular because all the spokes are in tension, but the weight supported at the hub increases the tension in the upper spokes and reduces it in the lower spokes (Figure 39).

The wire-spoked tension wheel was much lighter than the old wooden wheel and, when equipped with solid rubber or cushion rubber tyres, could absorb road vibrations. In cross section the spokes are arranged as in Figure 40, forming a triangulated structure able to sustain sideways forces on the wheel.

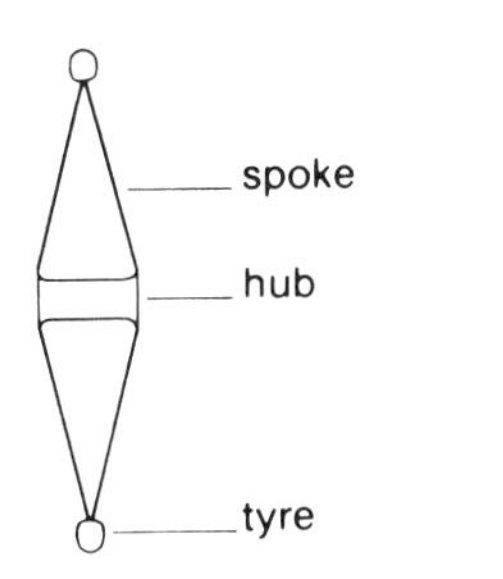

FIGURE 40
CROSS SECTION THROUGH WIRE-SPOKED WHEEL

The drawback of the tension wheel with spokes arranged radially, as in Figure 39, is that, when used as a driving wheel, it is not well able to transmit the torque (turning force) exerted by the pedals (or the chain drive), from the hub to the wheel rim. This is because all the forces act radially through the centre point of the hub (Figure 41), and there are no forces to prevent the hub rotating relative to the wheel rim when a torque is applied.

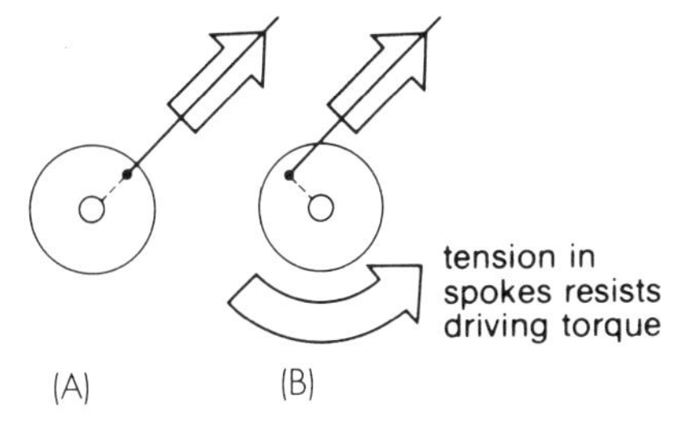

FIGURE 41

The weakness of radial-spoked wheels was overcome when James Starley patented the tangent-spoked wheel in 1874. In a tangent wheel the spokes are arranged at an angle to the hub and fixed a small distance from its centre (Figure 42). The tension in each spoke thus exerts a turning force on the hub that can resist an external driving or braking torque. The wheel as a whole is kept in equilibrium because for each spoke laid in the 'forward' direction there is another spoke balancing it laid in the 'backward' direction. The tangent spoked wheel therefore forms a structure braced against the torque exerted by either acceleration or braking. It supports the weight of the cycle and rider in the same way as a radial-spoked wheel.

What type of thinking do you suppose makes it possible for someone to conceive of a cycle wheel in which the spokes are in tension when previously wheels were cart-type with the spokes in compression?

Although I have no record of how Cowper conceived the idea of the wire-spoked wheel or how Starley came up with tangential spokes, it is fairly certain that they could not have done so without a theoretical, as well as a practical, understanding of structures and the principles of pre-stressing, triangulation and so on. In other words this is is a case of invention and design being dependent on the creative application of engineering science rather than intuition or craft knowledge.

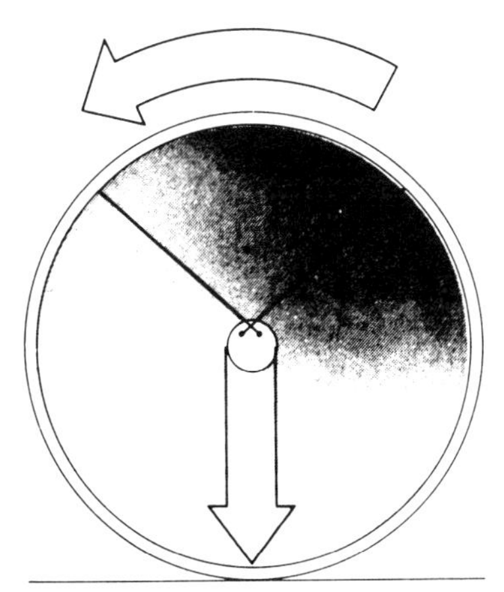

FIGURE 42
TANGENT-SPOKED WHEEL

The tangent-spoked wheel formed the definitive solution to the problem of providing light and rigid cycle wheels, and was rapidly adopted by manufacturers after its introduction. Radial-spoked wheels continued in use for a time and are still sometimes used today for the front (non-driving) wheels of racing bicycles in order to save weight, because they require fewer spokes.

FIGURE 43
ARIEL BICYCLE OF 1870, FORERUNNER OF THE ORDINARY OR 'PENNY FARTHING' BICYCLE

FIGURE 44
TYPICAL ORDINARY OR HIGH-WHEEL BICYCLE OF THE 1870S. FRONT WHEEL SIZES RANGED FROM 40 TO 60 INCHES

It is only since about 1980 that the wire-spoked cycle wheel has had any competition from new designs made possible by new materials; for example wheels with spokes made of composite materials and solid disc wheels made of carbon fibre (see Figure 23).

The Ordinary or high-wheel bicycle

The next major step in cycle evolution occurred in 1870 when James Starley and William Hillman patented the Ariel Bicycle (Figure 43). This was the prototype form of the Ordinary or high-wheel bicycle, later dubbed the 'penny-farthing'.

> What important innovations were incorporated in the design of the Ariel bicycle? (Look at Figure 43.)
>
> ---
>
> Wire-spoked wheels kept tensioned by means of an adjustable cross-bar fixed to the hub. The spokes were arranged like strings looped between the hub and wheel rim. To tension them the cross-bar was adjusted with wing-nuts, thus turning the hub relative to the wheel rim.
>
> A large front wheel and small rear wheel, which allowed the rider with each turn of the pedals to drive the cycle much farther and therefore faster than the old bone-shaker. Solid rubber tyres also made the ride smoother.
>
> Light all-metal construction (but still with a solid frame).

Later versions of the Ordinary bicycle simply took the concept to its logical conclusion – the front wheel reached the greatest possible diameter for direct drive, dictated by the rider's leg length, while the rear wheel shrank in size (Figure 44). The growing use of the high-wheeler in the 1870s and early 1880s for both sport and transport made inventors and manufacturers very active; numerous ideas for improvements were proposed, developed and adopted.

Among the most significant innovations that were applied to the high-wheeler as it evolved and improved were: the tangent-spoked wheel (discussed above); ball-bearings on the wheel spindles; and lightweight frames and forks constructed from hollow tubing. These improvements were of great significance because they were also applied to all succeeding types of cycle, including the Rover Safety.

> What were the main disadvantages of the Ordinary or high-wheel bicycle?
>
> ---
>
> Danger: the tendency to tip the rider over the front handlebars (a 'header') when the front wheel hit an obstacle.
>
> Skill: the agility and skill required to mount and ride the machine limited it to young and active riders.
>
> Inconvenience: it was necessary to dismount if the cyclist wished to stop for any reason. There was no free-wheel mechanism for coasting.
>
> Non-adjustability: many sizes were required to suit the leg lengths of different riders (although adjustable cranks provided some flexibility).

Although its advantages of speed, smoothness and simplicity gained the Ordinary bicycle much popularity, its inherent drawbacks – especially the danger of severe injury or death due to 'headers' – stimulated attempts to design *safer* types of cycle.

FIGURE 45
KANGAROO SAFETY BICYCLE PATENTED BY WILLIAM HILLMAN IN 1883

There were many approaches to the problem of designing safer cycles in the 1870s and 1880s but, as shown on the video, essentially these were of three types.

Safety Ordinaries

The least radical ideas for safety cycles involved making the front wheel of the Ordinary smaller, moving the saddle back and driving the front wheel indirectly via various mechanisms. The first successful safety Ordinary was the 'Xtraordinary', patented by Singer in 1878, in which a system of levers was used. The Xtraordinary also introduced the design principle of raked front forks, later adopted on the Rover Safety, which gave the bicycle self-centring steering characteristics.

Another important design, shown on the video, was the 'Kangaroo', invented in 1883 (Figure 45), which successfully established the idea of gearing up using a chain drive. By one revolution of the pedals the driving wheel travelled further than the wheel of an Ordinary cycle with direct drive. This principle was crucial to the development of the Rover Safety.

Tricycles, etc.

The second approach to safer cycles was the development of tricycles and other multi-wheel designs of many types and configurations. Although tricycles became very popular from the late 1870s to the 1890s, their evolution in the end turned out to be more important in stimulating the development of new components and mechanisms than as a rival to the bicycle. For example, the Salvo tricycle patented by James Starley in 1877 (Figure 46) was the first successful vehicle incorporating a roller chain drive, the forerunner of the bush roller chain invented by Hans Renold in 1880 (Figure 47). The development of an efficient and robust chain drive was a crucial step towards the evolution of the Rover Safety.

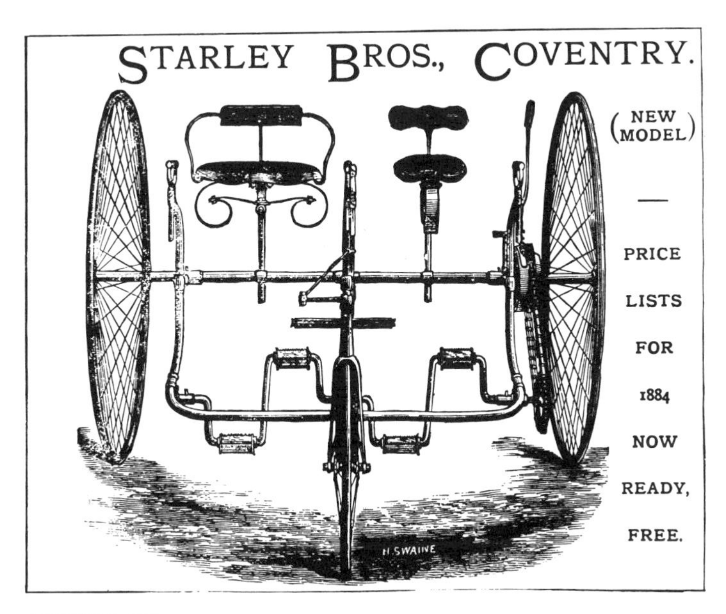

FIGURE 46
SALVO TRICYCLE PATENTED BY JAMES STARLEY IN 1887

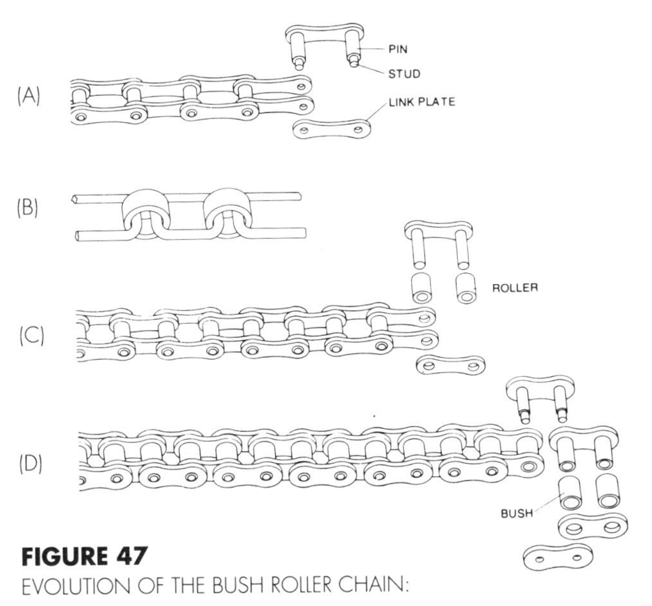

FIGURE 47
EVOLUTION OF THE BUSH ROLLER CHAIN:

(A) EARLY PIN OR STUD CHAIN

(B) MORGAN CHAIN

(C) ROLLER CHAIN

(D) BUSH ROLLER CHAIN INVENTED BY HANS RENOLD IN 1880 WHICH ELIMINATED THE HEAVY WEAR ON THE STUDS AND LINKS

FIGURE 48
H.J. LAWSON'S BICYCLETTE OF 1879 HAD REAR CHAIN DRIVE BUT RETAINED THE LARGE FRONT WHEEL

Chain-driven bicycles

The third direction of development in the search for safer cycles was the development of rear-driven 'dwarf' safety bicycles. Safety bicycles employing a chain drive to the rear wheel were built in England in the 1870s by several manufacturers. The 'Bicyclette' patented in 1879 (Figure 48) was one of H. J. Lawson's designs, but cyclists of the day ridiculed it, dubbing it 'the crocodile', and it did not meet with commercial success. A similar fate befell the 'Bate Shadow' chain-driven bicycle shown on the video. (You may recall from Block 1 the importance of getting the psychological aspects of a new design right if it is to gain market acceptance.) The designers of both machines seemed unable to break away from the idea of having one large and one small wheel, although with chain-geared drive this was not really necessary. Like many inventions, the rear chain-driven bicycle required someone else to develop the concept further in order to overcome the obstacles to change and to produce a successful innovation.

FIGURE 49
PIONEER CROSS-FRAME SAFETY HAD DIRECT STEERING AND GEARED-UP REAR CHAIN DRIVE

The first step was to shorten the frame, to employ direct steering to the front wheel and a geared-up drive to the rear wheel. This led to various designs of cross-frame safety bicycle, which were produced from 1884 (Figure 49). The next step was to combine these design features with a diamond-shaped frame. The final step was to produce a bicycle with a diamond-shaped frame and medium-sized wheels of similar diameter. The successful innovator was John Kemp Starley, the nephew of the prolific cycle inventor James Starley, with his Rover Safety bicycle.

INVENTION AND DESIGN OF THE ROVER SAFETY

I have followed the predecessors of the Rover Safety in some detail to show that it did not emerge in one giant leap 'out of the blue', but that it was the outcome of nearly twenty years of continuous cycle design evolution and component innovation. Like virtually all innovative designers, in creating the Rover Safety John Kemp Starley put together in a new way what had gone before.

What then was the nature of the thinking that led J. K. Starley to design the Rover Safety? Perhaps the best I can do to give you some feel for this is to quote directly from a retrospective lecture that Starley gave to the Society of Arts in 1898:

> A large number of different patterns were manufactured about this time [...]; but the year 1884 was a Kangaroo year. Records were made upon this machine which proved its superiority in speed over any bicycle previously introduced [...]. Although I was urged on many occasions to make such a bicycle, I felt the time had arrived for solving the problem of the cycle [...]. I therefore turned my attention solely to the perfection and manufacture of the Rover bicycle.
>
> The main principles which guided me in making this machine were to place the rider at the proper distance from the ground; to connect the cranks with the driving wheel in such a way that the gearing could be varied as desired; to place the seat in the right position in relation to the pedals, and constructed so that the saddle could be either laterally or vertically adjusted at will; to place the handles in such a position in relation to the seat that the rider could exert the greatest force upon the pedals with the least amount of fatigue; and to make them adjustable also. In the Rover, these, my cardinal principles, were all carried out, and although many alterations and improvements have been since effected in detail, it is a fact that there is not more than two or three inches difference today in any of the points first embodied in the Rover. [Figure 50]
>
> (Starley, 1898, p. 608)

FIGURE 50
CONFIGURATION OF THE 1885 ROVER SAFETY

FIGURE 51
LADDER ANALOGY USED BY STARLEY TO ESTABLISH THE CONFIGURATION OF THE ROVER SAFETY TO PERMIT OPTIMUM USE OF THE RIDER'S MUSCLE POWER

We see here some of the essential features of creative design thinking. Firstly, a dissatisfaction with what already existed. Secondly, some objectives on which to base an improved design.

Starley went on in his lecture to explain how he achieved his design objectives:

> The illustration which I now give [Figure 51] will show what was running in my mind at this time. I had been considering what a man could be compared to pedalling a bicycle, and the conclusion I had formed was, it largely resembled walking up a ladder, but with this difference, that whereas the pedals went down in the former, the man went up in the latter. I therefore had to determine where the handles should be placed to enable him to bring the whole of his weight on to the pedals, and, I think, the illustration will show my selection was correct. It was this question of the handle-bar which compelled me to adopt the present form of machine, as I could not get it sufficiently forward by the other type. It will be seen by the position of the handle-bar on the ordinary [penny-farthing] bicycle, that it was utterly useless and imperfect for this purpose.
>
> (Starley, 1898, p. 610)

FIGURE 52
FIRST PROTOTYPE OF THE ROVER SAFETY BICYCLE WITH A BRACED SINGLE-BACKBONE FRAME, BUILT IN 1884

You will notice that Starley achieved his ergonomically improved design configuration by means of an *analogy*. Analogy as you have seen is often the key to invention and creative design.

The first prototype of the Rover Safety was built and tested in 1884. It had indirect steering and a 36-inch front wheel (Figure 52). This clearly was not the elegant and efficient design that Starley wanted, and it was not until the third prototype of 1885 (Figure 53) that a diamond-shaped frame and direct steering to sloping front forks appeared. These were ideas originally developed by Humber for his 'Cripper' tricycle and safety bicycle of 1884. Like many creative individuals, Starley was willing to *adapt* solutions produced by others to develop his own designs.

FIGURE 53
THIRD PROTOTYPE OF THE ROVER SAFETY BICYCLE WITH A DIAMOND FRAME, BUILT IN 1885

By 1885, therefore, the essential form and most of the component parts of the modern bicycle had been established. It still remained to develop a triangulated frame structure by adopting straight tubes and adding a seat tube, and to curve the front forks at their ends in order to lighten the steering while retaining its self-centring characteristics. These improvements did not take long to appear. An early example that incorporated these design features was the Swift bicycle of 1893 shown on the video. By the mid-1890s most cycles being made were of this standard design configuration (Figure 54), and further improvements were adopted, such as the free-wheel mechanism, which allowed the rider to coast without taking the feet off the pedals.

FIGURE 54
FULLY EVOLVED ROVER SAFETY BICYCLE OF THE MID 1890S

INNOVATION AND DIFFUSION OF THE DIAMOND-FRAME BICYCLE

You have seen that, to be successful in innovation, inventiveness is not enough. What made the diamond-frame safety bicycle such an enormously successful innovation while earlier machines had only short-lived success? The answer is that this particular design configuration quickly proved itself to be superior to all the other types of cycle then prevalent: in speed, comfort, convenience of use, and frame strength and stiffness. And, after some of the strange machines that preceded it, the diamond-frame design almost matched the elegance and simplicity of the high-wheeler. Nevertheless, as Starley comments, his original Rover Safety had first to overcome the reaction of cyclists against anything new:

> At the time this machine was introduced, it created a certain amount of ridicule. It was so different in design from anything else made. It was not long, however, before we determined to prove that it possessed qualities which had never before been embodied in a cycle. In September, 1885, we broke the world's [cycle] record for one hundred miles on the road. From that day to this we have had with it one uninterrupted and ever-increasing success; while we have the satisfaction of seeing that all cycles [...] , whether driven by the front or back, or by the side wheels, have slowly but surely had to give way before it [...].
>
> (Starley, 1898, p. 608)

The rapidity with which the diamond-frame safety bicycle was adopted by manufacturers in Britain, Europe and America and the large numbers of safety bicycles that were sold in the period after 1885, depended not only on the superiority of the design but also on the social and commercial conditions towards the end of the nineteenth century.

The safety bicycle, especially after 1890, when it began to be equipped with pneumatic tyres, became a very efficient and comfortable machine, which created a cycling boom. In the mid-1890s the craze for cycling spread through the middle and upper classes, and to both men and women. To meet the demand, the number of cycle firms grew rapidly, until by 1897 there were over 830 manufacturers in the main industrial towns of Britain alone.

The cycling boom also led to important innovations in *manufacturing* technology:

> The arrival of the safety bicycle and the pneumatic tyre created such a boom that big makers in Coventry and Nottingham were forced to reorganise their methods completely to meet the demand ... Vast capacity was created by the large cycle makers and rows of the best machinery installed as a result of the heavy investment of the 1890s.
>
> (Saul, 1970, p.163)

By 1898 however the cycling craze was over, and the industry was left with a production over-capacity. The immediate result was that prices slumped, a large number of British manufacturers were forced into liquidation and many mergers took place. Major firms like Raleigh survived by adopting further manufacturing innovations to reduce costs while maintaining quality. The Rover Cycle Company founded by J. K. Starley also survived but, like many other famous cycle manufacturers (e.g. Humber, Hillman, Singer, BSA), diversified into developing and making motor cycles and motor cars.

The first vehicles with internal combustion engines were motorised bicycles and tricycles (Figures 55 and 56), which appeared in Germany in the same year (1885) that the bicycle had evolved to the stage of the Rover Safety. The cycle industry and many of the components and manufacturing innovations it produced thus provided the basis for the newly emerging motor industry. In Block 5 we look at the motor industry, and in particular how the design of cars is dependent on manufacturing technology.

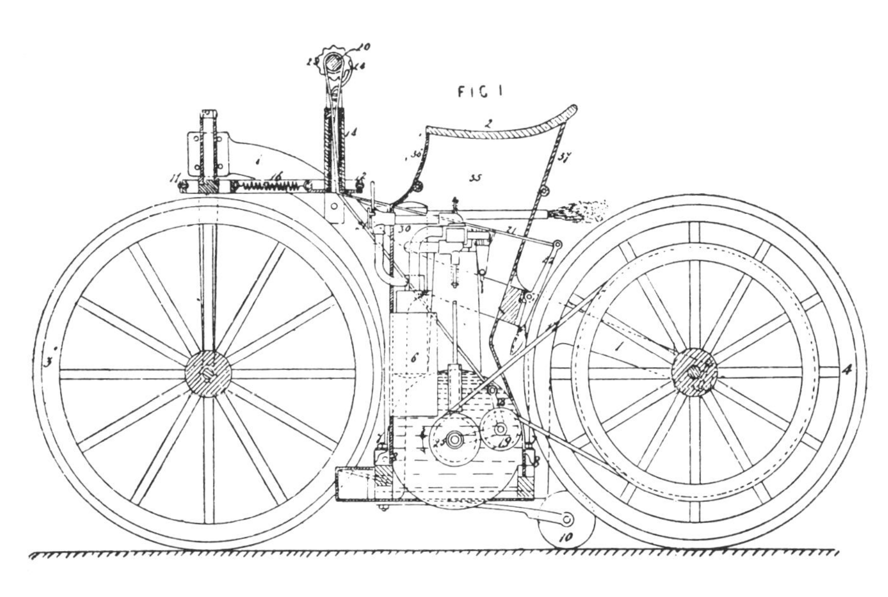

FIGURE 55
DAIMLER MOTOR BICYCLE, THE FORERUNNER OF THE MODERN MOTORCYCLE WAS DESIGNED AND BUILT BY GOTTLIEB DAIMLER IN 1885. THE VEHICLE WAS EQUIPPED WITH A SINGLE-CYLINDER INTERNAL COMBUSTION ENGINE. THE TWO SMALLER JOCKEY WHEELS RETRACTED WHEN THE MACHINE WAS UNDER WAY

FIGURE 56
BENZ MOTOR TRICYCLE, THE FORERUNNER OF THE MODERN AUTOMOBILE, ALSO MADE ITS APPEARANCE IN 1885. THIS FIRST DESIGN OF CARL BENZ HAD A ONE-QUARTER HORSEPOWER INTERNAL COMBUSTION ENGINE AND ELECTRIC IGNITION. BY 1886 IT COULD TRAVEL RELIABLY AT 9 MPH. (THE PHOTOGRAPH IS OF A REPLICA OF THE ORIGINAL VEHICLE)

SAQ 9
What was J.K. Starley's main contribution in designing the Rover Safety bicycle? On what previous innovations did he rely in producing the design?

4.2 THE PNEUMATIC TYRE

At the time of the Rover Safety the most important component innovations that remained to be successfully applied to the bicycle were: the pneumatic tyre, patented by Dunlop in 1888; the Bowden cable and rim brake introduced after 1900; the epicyclic hub gear, patented by Sturmey and Archer in 1902; and the derailleur gear introduced in France around 1909.

In this second example I am going to examine the pneumatic tyre. The pneumatic tyre was probably the most significant of these remaining developments since, by greatly improving the comfort and speed of the bicycle, it helped to create the cycling boom of the 1890s, and later provided the basis of a major industry producing tyres for vehicles of all kinds.

Just as the diamond-frame bicycle displaced the various designs that preceded it, the pneumatic cycle tyre displaced the solid rubber tyres and cushion tyres with which cycles had been equipped prior to its introduction. The main reason for the superiority of the pneumatic tyre over solid or cushion tyres was that, on ordinary roads with irregular surfaces, the pneumatic tyre not only increased the rider's comfort by absorbing the bumps, but reduced the effort required for cycling by minimising the energy losses due to jarring of the unsprung mass of the cycle and rider as they travelled over any irregularities.

ORIGINS OF THE PNEUMATIC TYRE

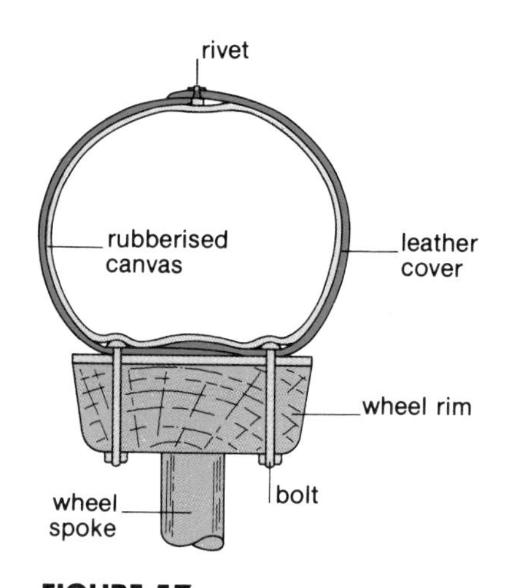

FIGURE 57
THE FIRST PNEUMATIC TYRE: THOMPSON'S AERIAL WHEEL OF 1845

The original inventor of the pneumatic tyre was a Scottish civil engineer, Robert Thompson, who patented the principle in 1845. He failed, however, to make a commercial success of his invention and his name is consequently almost unknown. The reasons for Thompson's failure to achieve an innovation are instructive. Neither the bicycle nor vulcanised rubber were developed at the time of Thompson's pneumatic tyre and so he had to think in terms of the vehicles and materials available to him. He envisaged that his 'aerial wheels' would be used on heavy vehicles such as horse-drawn wagons, steam vehicles and even railway carriages. He constructed his tyres from a tube of rubberised canvas with a leather cover that was bolted to the wheel rim (Figure 57). Although he demonstrated that tractive effort on rough roads could be considerably reduced with the aid of his aerial wheels, there was never a sufficiently large market in pneumatic tyres for heavy vehicles for his invention to be commercially viable. It was Thompson's misfortune to have been too far ahead of his time.

DUNLOP'S PNEUMATIC CYCLE TYRE

The second inventor of the pneumatic tyre was another Scotsman, John Boyd Dunlop, who (invalidly) patented the idea in 1888 while practising as a veterinary surgeon in Belfast. Although Dunlop's patent only envisaged the application of pneumatic tyres to light vehicles, the commercial success of his invention allowed it to be developed and applied to all forms of road vehicle.

Because of the disputed patent, the precise details of the origins of Dunlop's pneumatic tyre are uncertain. Arthur du Cros (1938), one of the founders of the business which became the Dunlop Tyre Company, provides one version, but according to *The History of the Pneumatic Tyre* by Dunlop himself, the following events were involved.

Dunlop's mind, like Thompson's and many others before him, had been considering the problem of vehicle vibration. For a period of over twenty years Dunlop had considered, and then abandoned as impractical, various forms of spring wheels, including the idea of a flexible steel band supported by rollers. Eventually he thought of the idea of a wheel with an air-filled tyre, which he envisaged as being made of cloth, rubber and wood. Once the idea of an air-filled tyre had occurred to Dunlop, he rapidly proceeded to conduct a rough experiment. For the experiment Dunlop used a solid disc of wood to which he nailed a pocket of linen that enclosed an inflated rubber tube. He compared the rolling properties of this wheel with a solid rubber-tyred wheel, with promising results: 'I first threw the solid tyred wheel along the yard, but it did not go the full length of it. Then I threw the air tyred disc. It went the whole length of the yard and rebounded with considerable force off the gate.' (Dunlop, 1922, p. 11.)

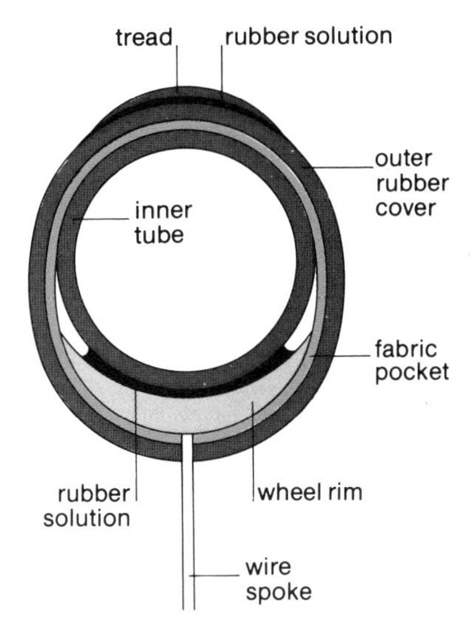

FIGURE 58
DUNLOP'S ORIGINAL 'MUMMY' PNEUMATIC TYRE OF 1888

Encouragement for the practical development of the air tyre came from Dunlop's son. The boy wanted to beat his friends in cycle races and also complained of the difficulty of cycling fast on the Belfast streets, which at the time were surfaced partly with granite setts.

To gratify his son's desire for speed, Dunlop's experimentation progressed to building up and fitting a pair of wheels with inflatable tyres to the boy's tricycle. This was not too difficult for Dunlop, who already had considerable experience of developing and making rubber appliances for his veterinary work. For this key experiment Dunlop made and used wooden-rimmed wheels with the rubber tube contained in a canvas pocket wired and glued to the rim. The canvas was protected by an outer wrapping of sheet rubber stuck on with rubber solution. A simple non-return valve in the inner tube allowed inflation, but no deflation. This was later known as the 'mummy' or 'rag' tyre, comprising a triple tube of rubber, canvas and rubber inflated with compressed air (Figure 58).

The prototype pneumatic-tyred tricycle was given a test run by Dunlop's son in February 1888. The trial showed that pneumatic tyres were workable and Dunlop proceeded to fit a second tricycle with his new tyres. This was ridden by a friend of Dunlop, who was immediately struck by the comfort, smooth running and speed of the machine and encouraged Dunlop to exploit his idea. In July 1888, with the help of his friend, Dunlop applied for a patent on his invention.

> Using the terms introduced in Section 3 to decribe the creative process, how might you explain the thinking by which Dunlop invented his pneumatic tyre?

> Dunlop had been considering the problem of vehicle vibration for a long time before he came up with the idea of the pneumatic tyre. He underwent a period of *preparation* and *incubation*. Although Dunlop does not say how he got the idea for an air-filled rubber tyre, it seems likely that it stemmed from his experience of making rubber appliances during his veterinary work. Another version is that he got the idea from animal intestines. His thinking therefore involved either *adaptation* or *biological analogy*. Once Dunlop had thought of an air-filled tyre he performed experiments to test and develop it. He undertook a process of *verification*.

The invention of the pneumatic tyre seems to follow closely the creative process outlined in Section 3.

INNOVATION AND DIFFUSION OF THE PNEUMATIC CYCLE TYRE

Dunlop attempted to interest manufacturers in his invention by introducing pneumatic-tyred bicycles into cycle racing. However, the sausage-like pneumatic tyre was at first greeted with scepticism among cyclists, and with outright ridicule from the public. It eventually gained acceptance when Irish cyclists using pneumatic-tyred machines began in 1889 and 1890 to win races against English riders of solid and cushion-tyred bicycles.

Among the earliest converts to pneumatic tyres was Arthur du Cros's father. Arthur and his brother were members of the Irish cycle racing team. Convinced of the commercial potential of the Dunlop tyre, the du Cros brothers proceeded to establish, in Belfast in 1889, the world's first company for manufacturing pneumatic tyres.

Despite the early racing successes, the pneumatic tyre still met with considerable opposition from cyclists and manufacturers. English manufacturers in particular were not willing to modify their cycle frames to accommodate the much fatter pneumatic tyres. Cyclists preferred the slimmer and more elegant-looking solid and cushion tyres. This led to heated discussions, arguments and controversies throughout the cycling world, which in the end, by publicising the pneumatic tyre (especially when races were won using it), helped to generate a public demand.

The Dunlop tyre of 1888, however, suffered from several deficiencies, the most severe of which was the fact that it punctured easily and was permanently fixed to the wheel. Repairing or replacing one of these early pneumatic tyres was a complex and skilled operation.

The pneumatic tyre therefore required the attention of other inventors before it was developed sufficiently to make its widespread adoption practical and acceptable. Various inventors devised methods for attaching the outer covers of pneumatic tyres to the rims of cycle wheels so that the whole tyre could be easily removed for repair or replacement. Of these methods two important ones were C. K. Welch's 1890 patent for 'wired-on' tyres (Figure 59(A)) and W. E. Bartlett's patent, also of 1890, that led to the 'beaded-edge' tyre (Figure 59(B)). In France, Edouard Michelin patented his design of detachable pneumatic tyre in 1891.

With the appearance of the detachable pneumatic tyre, plus several other improvements in tyre design, construction and manufacture, the pneumatic cycle tyre rapidly gained acceptance after 1892 and by the mid-1890s was in general use.

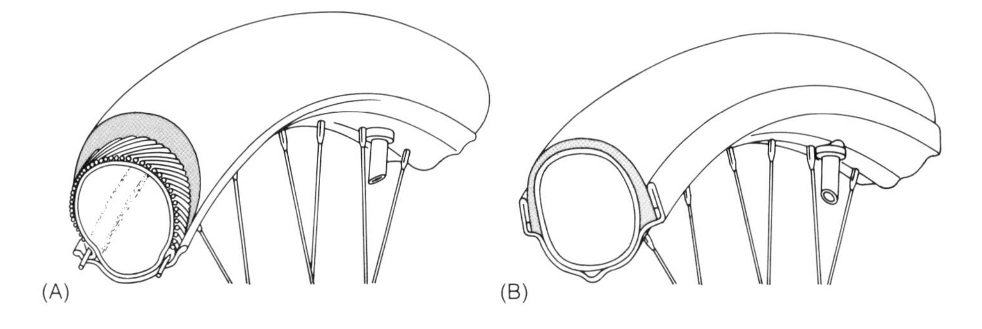

FIGURE 59
(A) WELSH-DUNLOP WIRED-ON DETACHABLE TYRE OF 1890
(B) BARTLETT-CLINCHER BEADED EDGE DETACHABLE TYRE OF 1890

SAQ 10
What general lessons about invention and innovation may be learnt from Thompson's unsuccessful and Dunlop's successful attempt to introduce the pneumatic tyre?

4.3 THE DURSLEY PEDERSEN BICYCLE

In the previous sections I have described two enormously successful innovations. In this section I want to look at a bicycle that proved to be much less successful as an innovation, even though it was in some ways superior in design to the diamond-frame safety. You will discover that ingenuity and technical sophistication in design are not enough, nor often even appropriate, for successful innovation. For commercial success, the price, the balance between the practical and the psychological aspects of the design, and the social and economic conditions have to be favourable to the innovation.

The example I will be using is known as the Dursley Pedersen bicycle, named after its Danish designer Mikael Pedersen and the Gloucestershire

FIGURE 60
DURSLEY PEDERSEN BICYCLE INTRODUCED IN 1897

town of Dursley, where Pedersen settled and began manufacturing his bicycle.

The Pedersen bicycle was probably invented in Denmark in the 1880s, but did not make its commercial appearance until 1897 and so enjoyed all the benefits of the pneumatic tyre, roller chain drive, ball-bearings, etc. The unique feature of the Dursley Pedersen was a frame made up of thin-walled tubes in a lightweight triangulated structure which supported an unusual hammock type seat (Figure 60).

In what follows I have drawn heavily on *The Ingenious Mr. Pedersen* by David Evans (1978).

INVENTION OF THE DURSLEY PEDERSEN BICYCLE

The impetus for invention and design, as you have seen, often stems from dissatisfaction with existing products. Pedersen's particular dislike was the saddle used on safety cycles. This led him to develop a novel bicycle designed on the 'space frame' principles used in bridge trusses and tower cranes.

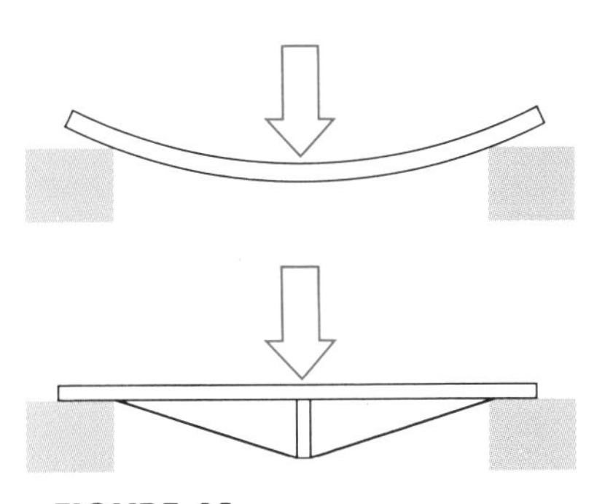

FIGURE 61
TRIANGULATED STRUCTURE

Each frame member on Pedersen's design consisted of pairs of thin tubes joined to make triangles which in turn were linked into a triangulated configuration. The result was a frame that was, in theory at least, very light and strong, since a structure that comprised members configured into triangles should only have to sustain compression or tension forces in its components (Figure 61).

Pedersen came to England and patented his bicycle in 1893. You can see part of the original patent specification illustrated in Figure 62 and a demonstration of the machine on the video.

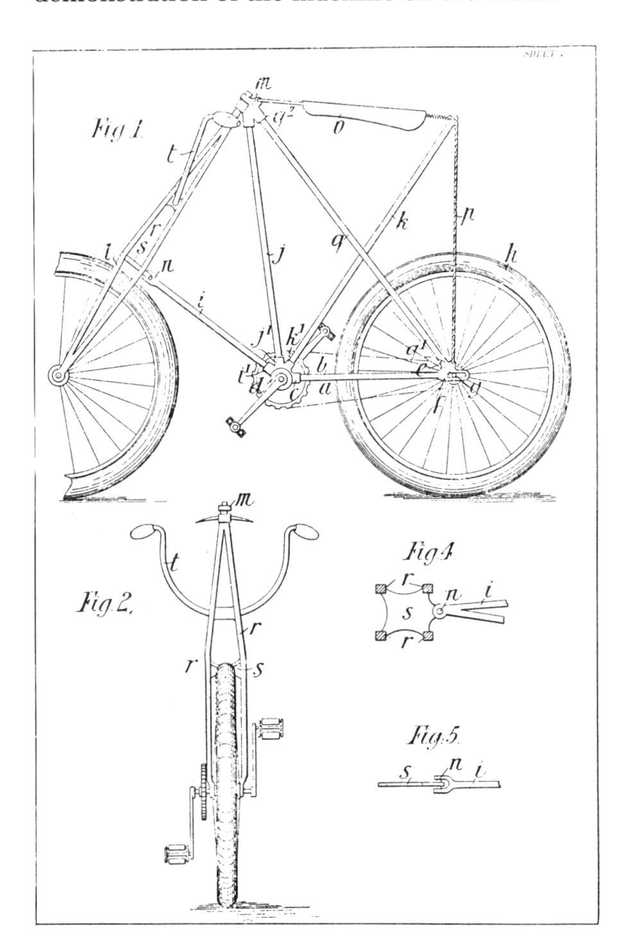

FIGURE 62
PROVISIONAL PATENT SPECIFICATION FOR PEDERSEN'S IMPROVED BICYCLE, 1893

An account of how Pedersen created his invention appears in one of the early catalogues of the company. I have taken a few extracts from that account to give you a feel for Pedersen's main lines of thinking.

> I have been a cyclist for more than twenty years, and have done much hard riding. [...] I soon found there was much room for improvements in the construction of cycles, although it was only when I got my first 'safety' that I saw how much yet remained to be done in this direction.
>
> The part of the machine in general use which I found especially imperfect was the seat. [...] I made several experiments before I could get exactly what I wanted, but finally my efforts were crowned with success. The seat which I have devised is [...] made of strings of different degrees of tension, running from a point in front to a cross steel bar giving the requisite width behind ... The seat is suspended between two supporting points about two feet from each other. [...]
>
> Seeing that I should want so much room between the two points for suspending the seat, I found it almost impossible to make a new seat frame which would not be too heavy, and which would have an elegant appearance. I resolved to make a cycle frame which would carry the seat without the necessity of having a special seat frame at all. In the ordinary 'safety' I found the frames were so far from perfection that I had to abandon the system entirely.
>
> (Evans, 1978, p. 107)

Pedersen goes on to describe how he built up a frame linking the seat supports, bottom bracket and wheels with a series of double triangles (e.g. the triangle formed by tubes bjq shown in Figure 62).

> How would you describe the thinking that led Pedersen to his novel bicycle design?

> What is interesting about Pedersen's account is the way in which his thinking seems to have been dominated, first by the need to design a frame around his unconventional hammock seat (the *primary generator* for the design), and then by the idea of elegant triangulated structures. Although the result is undoubtedly ingenious, it was clearly based more on intuition than on an understanding of the various dynamic loads that act on a cycle when in use.

Early Dursley Pedersen models weighed about 9.1 kg (20 lb) and cost about £25, making them equivalent in both weight and price to top-class racing bicycles of today. But some incredibly light machines were made by Pedersen, one (without brakes) weighing only 4.1 kg (9 lb), less than even the lightest modern bicycles.

INNOVATION AND DECLINE OF THE DURSLEY PEDERSEN

The Dursley Pedersen bicycle was first introduced on to the British market in 1897. Sales were slow at first, but were good from 1905 to 1908, and then fell slowly until 1914, when production ceased. The company making Dursley Pedersens was wound up in 1917, although commercial manufacture on a small scale continued into the 1920s.

Why did the structurally sophisticated Dursley Pedersen eventually fail while the conventional design flourished? The reasons are technical, commercial and social. Let's start with the technical reasons.

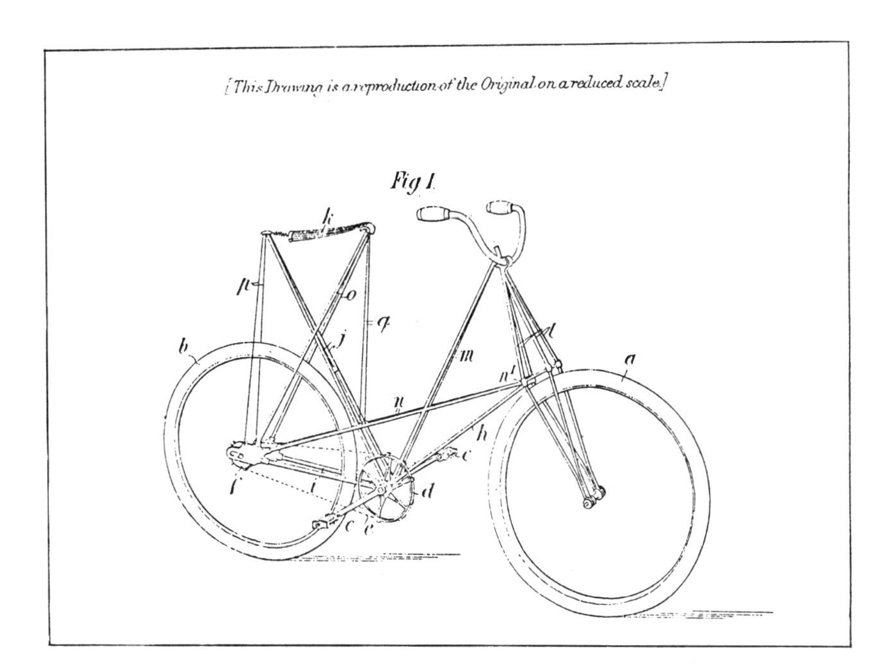

FIGURE 63
FULLY TRIANGULATED WOMAN'S FRAME BICYCLE PATENTED BY PEDERSEN IN 1899

Look carefully at the drawings of the Dursley Pedersen bicycle in Figures 60, 62 and 63. Compare these with an illustration of the conventional diamond-frame bicycle (e.g. that shown in Figure 54). Can you deduce some of the reasons why Pedersen's design failed to replace the conventional design?

This is what I observed:

The general visual impression of the Dursley Pedersen is one of complexity when compared to the conventional design. This has two implications. First, this complexity does not appear to offer major advantages over the conventional design. The hammock saddle, which inspired Pedersen, could have been provided on a simpler frame (e.g. a modified conventional women's frame). Secondly, manufacturing a complex frame like that on the Dursley Pedersen is bound to be more expensive than making the conventional diamond frame.

While the thin tubing gives an impression of lightness, at the same time the structure appears delicate and easily damaged and could lack stiffness to the twisting forces exerted by pedalling. In fact cyclists who raced early Pedersens complained of the flexing of the bottom bracket (bearing the pedals) in hard riding. The weight of Dursley Pedersens increased gradually from 9.1 kg (20 lb) in 1897 to 16 kg (35 lb) by 1914, presumably due to the use of thicker tubing. The weight advantage over conventional machines was therefore lost.

The apex of the frame, where on the man's model the front of the saddle is attached, looks as if it could be a hazard in an accident. It is interesting therefore that later models appear to have had their handlebars mounted in a safer position, at the top of the forks.

What of the other reasons for the eventual disappearance of the Dursley Pedersen?

Probably the most obvious was that, just at the time of its introduction, the bicycle craze of 1895–7 was over and prices were in decline. The

FIGURE 64
FOLDING DURSLEY PEDERSEN BICYCLE

complexity of manufacture meant that Dursley Pedersen bicycles remained twice as expensive as standard conventional machines; they were firmly in the shrinking 'de luxe' cycle market. A folding version was produced for military use (see Figure 64 and Video 3, Section 3, 'The challenge of the portable bike'). But while Dursley Pedersens initially offered a weight advantage and won several races, this was only sufficient to attract those cycling enthusiasts who were not deterred by their unconventional appearance. As you have seen – and will see again in Section 6 – visual appearance is very important to success in bicycle design and innovation.

Pedersen's inventive, but stubborn and often unbusinesslike, character also contributed to the commercial problems of the Pedersen Cycle Company. In 1902 he patented a three-speed hub gear using the counter-shaft principle (the epicyclic hub gear of Sturmey and Archer was being developed at the same time). Pedersen's gear, although sound in principle, proved unreliable in use because of a friction clutch. Pedersen refused to alter the clutch, instead he introduced minor design changes which increased manufacturing costs. The unreliable gear damaged the reputation of the Dursley Pedersen Company at a time when bicycle sales were falling, and in 1904 the company was taken over by another firm. Under the new management the friction clutch was quickly replaced by a toothed drive, which worked, and the cycle business was wound up. Like many inventors, Pedersen in the end was left disillusioned and virtually penniless.

Pedersen's design ideas have however been revived by a Danish blacksmith who, in 1978, started manufacturing a modern version of the Pedersen bicycle. And in the mid-1980s an improved Pedersen was made in Britain by a firm located a few miles from Dursley.

SAQ 11

What were the main sources of the ideas underlying the conception of the Rover Safety bicycle, Dunlop's pneumatic tyre and the Dursley Pedersen bicycle?

SAQ 12

On the basis of the three examples in this section, identify conditions for commercial success in innovative product design.

5 THE INVENTOR AND CREATIVE DESIGNER

In previous sections we looked at how some specific new inventions and design ideas came about. In doing this I touched upon the characteristics of some of the inventors and designers involved. In this section I want to look more closely at the characteristics of individuals noted for their creativity abilities, and to help you to discover the extent to which you share those characteristics. I hope to show that everyone has the potential for generating original and useful ideas – and that includes you, even if you do not consider yourself to be particularly creative. Some people of course are naturally creative, or have become so through experience, while others might need some guidance and techniques to bring out their latent creative abilities. These approaches and aids to creative thought will be the subject of Section 7.

This section involves use of the audio-cassette for this Block. You will also need your *Audio 2 Study Guide* to complete Section 5.5.

There have been numerous studies of creative individuals and there is quite wide agreement on their characteristics. You should already have some idea of some of these characteristics from your study of the Block so far. For example, the description in Section 4 of J.K. Starley's innovative work in bicycle design demonstrates a creative individual's ability to recognise a worthwhile problem and desire to solve it; their willingness to question established dogma; their use of knowledge and experience to come up with viable solutions, and their dedication, which is needed to overcome the obstacles to change.

Some of these characteristics are to do with the *personality* of the individual, others refer to particular *intellectual* abilities and *practical* skills, and yet others concern the *knowledge* possessed by the creator. Some of the characteristics are relevant to all types of creative activity, while others are specific to invention and creative design. Let us look at these in turn.

5.1 THE CREATIVE PERSONALITY

Since there is not much that can be done about one's basic personality I shall not say much about this, except that various studies have shown that highly creative individuals, in all fields of endevour, tend to share certain personality traits – although of course no creative person is entirely like any other.

For example, creative individuals tend to be perceptive and sensitive to their environment and to problems and opportunities around them. Where others may be content to tolerate a difficulty or be satisfied with the status quo, the creative individual, especially in the fields of science, technology and design, is usually on the lookout for problems, improvements and solutions. This has been described as 'constructive discontent' (a term introduced in Block 2) to indicate that creative designers or inventors are frequently dissatisfied with what exists, but do not just moan about it, they want to do something to make the situation better. Where others might give up when solving a problem gets difficult and ideas fail, creative individuals will generally persist until a solution is found – even if it takes years of effort. This willingness of creative individuals to persist against the odds usually comes from their intense interest in a particular task or problem and from the deep satisfaction they get from solving it. Creativity, perhaps more than anything else, comes from individuals' motivation and enjoyment of what they are doing.

Can you give an example from a previous section of this Block text of an inventor or designer who displayed at least some of the characteristics described above?

Although I said little about his personality, Chester Carlson, the inventor of xerography, clearly displayed 'constructive discontent', plus the persistence to see a technically very difficult problem through to a solution. He worked alone or with one assistant, rather than in a company or research institution, at least when producing the basic invention.

Another example is Mikael Pederson, designer of the Dursley Pederson bicycle, who also showed constructive discontent, but whose stubborn personality and unwillingness to leave successful designs alone was a factor in the subsequent failure of his cycle business. This indicates that inventors and creative designers, although good at producing new ideas, may not always be well-suited to other tasks, such as running the business that may result from that idea.

5.2 CREATIVE ABILITIES AND SKILLS

If we turn from general personality traits to the more specific abilities and skills required for creative thinking and problem solving the picture becomes more complex.

CONVERGENT AND DIVERGENT THINKING

In the 1960s and 1970s the relationships between creativity and different aspects of intelligence were studied in detail by J.P. Guilford, an American psychologist. Guilford showed that creative abilities are part of overall intelligence but are poorly related to the mental abilities usually measured by intelligence tests.

Conventional IQ tests tend to measure an individual's ability to produce correct or expected responses to problems and puzzles, such as those shown in Figure 65.

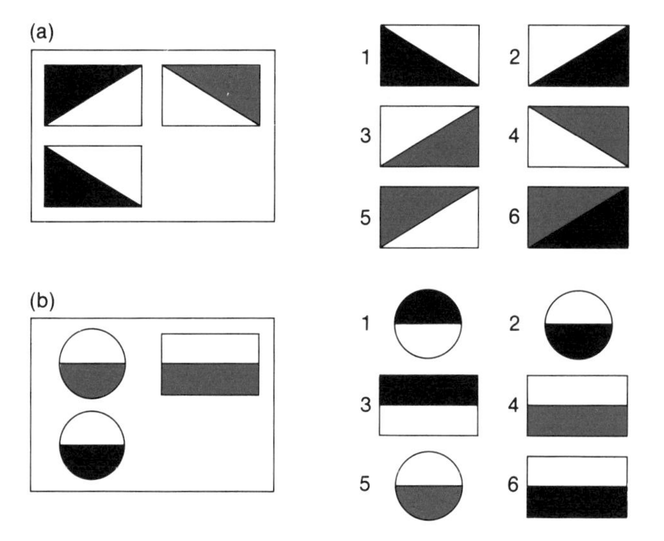

FIGURE 65
WHICH SHAPE (1–6) FITS THE SPACE? TYPICAL IQ TEST QUESTIONS REQUIRING LOGICAL, CONVERGENT THINKING

What such tests measure is the ability to analyse information in order to select an answer from alternatives, hence the term **convergent thinking.** Guilford argued that creative problem solving was more related to another mental ability, namely **divergent thinking.** Furthermore he concluded that divergent thinking in creative problem solving comprised a number of separate abilities, which he called fluency, flexibility and originality.

Fluency is the ability to make use of stored information to rapidly produce many different responses to a given problem. For example, here is a famous test of what Guilford called 'ideational fluency':

> Write down as many uses as you can think of for a brick. Allow yourself two minutes. Stop reading and try it yourself.

> How many ideas did you come up with?
>
> Here are two sets of responses to this question from sixth-form schoolboys given by Hudson (1967).
>
> *Schoolboy A:* For building, for throwing through window.
>
> *Schoolboy B:* To break window for robbery, to determine depth of wells, to use as ammunition, as pendulum, to practise carving, wall building, to demonstrate Archimedes' Principle, as part of abstract sculpture, cosh, ballast, weight for dropping thing in river, etc., as a hammer, keep door open, footwiper, use as rubble for path filling, chock, weight on scale, to prop up wobbly table, paperweight, as firehearth, to block up rabbit hole.

The second individual with his twenty-one ideas clearly shows greater fluency of thinking than the first, who produced only two. Design students on average produce between ten and twenty ideas in this exercise. How did you do? If you managed less than five ideas you may not be a naturally very fluent thinker, and could benefit from trying some of the idea generation techniques in Section 7.

Flexibility is the ability of an individual to alter their mental set when approaching a problem and to produce ideas out of the usual categories. The second set of responses to the 'uses of a brick' exercise shows more flexibility as well as fluency. Ideas such as using the brick as a pendulum and as part of an abstract sculpture show flexibility because they break away from the usual view of bricks being used for building. This ability of an individual to 'roam about' in his or her thinking, even when it is not necessary to do so, Guilford called 'spontaneous flexibility'. A related skill, termed 'adaptive flexibility', is a person's ability to solve problems that require breaking away from common assumptions and preconceptions. The nine dot puzzle (Figure 66) is a well-known problem whose solution requires adaptive flexibility. It is one of the problems you are set in the 'Thinking styles exercises' later in Section 5.5.

FIGURE 66
NINE DOT PUZZLE. JOIN ALL NINE DOTS USING FOUR (OR FEWER) CONNECTED STRAIGHT LINES WITHOUT TAKING THE PEN FROM THE PAPER OR GOING BACK OVER A LINE

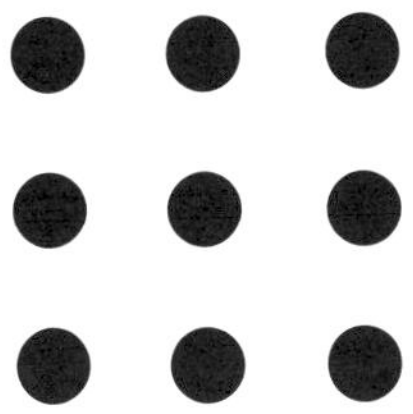

Here is another problem for you to try, taken from James Adams' book *Conceptual Blockbusting* (Adams, 1987, p. 54):

> Assume that a steel pipe is embedded in the concrete floor of a bare room as shown [Figure 67]. The inside diameter is 0.06 in larger than the diameter of a ping-pong ball (1.50 in) that is resting gently at the bottom of the pipe. You are one of six people in the room, along with the following objects:
>
> 100 ft of clothes-line
>
> A carpenter's hammer
>
> A chisel
>
> A file
>
> A wire coat hanger
>
> A monkey wrench
>
> A light bulb
>
> List as many ways as you can think of (in five minutes) to get the ball out of the pipe without damaging the ball, tube or floor.'

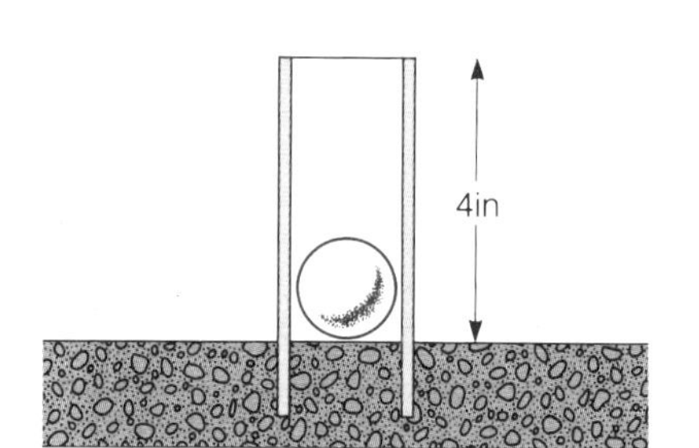

FIGURE 67
BALL AND PIPE PROBLEM

Stop and attempt this problem now.

Adams notes that if you produced a long list of different methods for retrieving the ball from the pipe, you are a fluent thinker. However, if in addition your thinking is *flexible* you would have produced a wide variety of different approaches to the problem. For example, filing the coat hanger in two, flattening the ends and making large tweezers to retrieve the ball is one fairly common solution. Smashing the handle of the hammer with the monkey wrench and using the resulting fragments to retrieve the ball demonstrates more flexibility of thought, since using a hammer in this way involves breaking out of one's usual patterns of thinking about tools.

Originality is the ability to generate novel, unusual or ingenious ideas in response to a problem. Did you think of getting the group to urinate into the pipe to get the ball to float up? If not, why not? Adams argues that original ideas like this are sometimes prevented by a variety of mental and other blocks, in this case a social taboo. Another version of the ball and pipe problem includes among the objects in the room a tray containing cups of coffee for all the participants. In this case someone usually thinks of the (rather less original) solution of pouring coffee into the pipe to make the ball float up.

The work of Guilford and others which stressed the importance of divergent thinking in creativity has been very influential. It is the basis of various 'creativity tests', and also underpins brainstorming and other creative problem solving techniques that are specifically designed to stimulate people to think of many possible ways of tackling a problem. Flexible and original thinking also has much in common with Edward de Bono's more widely-known concept of **lateral thinking.** However, it is a mistake to simply equate divergent or lateral thinking with the ability to produce creative ideas. It is perfectly possible to produce lots of highly divergent ideas none of which are of any practical or other value. In many fields, including invention and design, convergent thinking is as important as divergent thinking in solving problems.

Michael French even argues that in engineering design:

> ... it would make much better sense to call creative processes convergent ... The designer is choosing one form from many, narrowing down to a particular solution ... one of the distinguishing features of the good designer is the ability to *converge from a wide base* on a good choice.
>
> (French, 1988, p. 277)

This of course begs the question of where the 'wide base' of ideas comes from in the first place. It could come from the designer's extensive knowledge and experience (see below), but it could equally well stem from a divergent search for ideas. What French probably means is that a creative designer must be able to choose fruitful solutions from the possibilities he or she might have considered either explicitly on paper (as in the car horn example of Section 1) or subconsciously within the mind. As you also saw in Section 1, creativity in design, as well as generating alternatives, involves *synthesis* of a wide range of information into a single solution.

In practice therefore it seems that *both* divergent and convergent thinking are important when creating a new product, device or system. Divergent thought is needed to generate original ideas, while convergent thinking is needed to select between them and to develop the chosen idea into a working practical design. The precise mix of divergent and convergent thinking required depends on the nature of the problem. Some design tasks are much more open-ended than others and therefore involve more divergent thought than highly constrained design work. The good designer is someone who has both sets of mental skills needed to produce creative work and can *switch* between the two – at one moment turning off the critical faculties to produce solution ideas and then subjecting them to rigorous evaluation and choice – many times over during the problem solving process.

VISUAL AND PRACTICAL SKILLS

What other abilities and skills are needed in invention and creative design? A good summary of some of the more important of these is provided by Professor Meredith Thring, a well-known inventor and engineering designer, in the following extracts from *How to Invent*:

> First he [Thring notes that he is using 'he' to include women] has to develop in himself the *inventor's eye*. This means that he learns to look at every operation or construction around him and think 'why is it done that way or built that way' and whether there is a better way of doing it. This applies to buildings, bridges, tools, kitchen utensils, factory machines ... to waterfalls, waves and winds as usable sources of energy. He also thinks about and experiments with every manual job he does, such as washing up, digging in the garden or sawing wood, to see if he can do it better, for example so as to be less tired, so as to use less water or electricity ... and to do it quicker ... This development of the inventor's eye not only teaches one to think back to first principles and to think what are the objectives of the designer, but it also stocks one's brains with all sorts of useful pieces of practical information.
>
> (Thring and Laithwaite, 1977, p. 44)

So one requirement of the inventor or creative designer is to have a questioning attitude and to be curious about why and how things work and might be improved – what I referred to earlier a constructive discontent.

Another crucial requirement is what Thring calls 'thinking with the hands':

> The intellectual brain can only deal with concepts, words, ideas and theories, whereas it is through the senses and the hands that one has experience of the behaviour of solids, fluids, flames, explosions, mechanisms, electric currents, magnetism, forces and heat ...
> As far as possible therefore one prepares oneself for the inventive act by playing with models, sketching diagrams and looking at the world of existing solutions to the problem.
>
> (pp. 90–91)

You may recall James Dyson on the video talking about his need to use and observe existing products and to engage in practical workshop activities in order to produce innovative designs.

A third major requirement, which I hinted at earlier, is the ability to 'suspend judgement' on new ideas and to record them in a sketch or other form before attempting to evaluate them:

> An essential characteristic of the creative state is that one's critical faculty which normally inhibits all new ideas from being formed is completely switched off. Thus ideas are born, many of which will be killed when the critical faculty is switched on again, but others can be followed up by further ideas which meet the criticisms if one is still in the creative state. Thus one has to learn the knack of switching off one's critical faculty at least till some promising idea is actualised by a sketch on the back of an envelope or a few words describing where the solution can be found.
> (p. 46)

Sketching and drawing are not just vital skills for *recording* inventions and creative design ideas, they are just as important in *exploring* possible solution concepts and developing them to the prototype stage. Thring says:

> Rough sketches are the first technique. The back of an envelope is the most expressive description of the medium used, but a pocket notebook is the inventor's usual companion, preferably of unlined blank paper, and the sketches are made very roughly so that one has no compunction in throwing dozens away while the idea is taking shape. The ability to draw freehand is just as essential a part of the inventor's tool-kit as it is for the lecturer or artist. Drawing with a soft pencil on unlined paper enables one to use a rubber frequently, which is a valuable aid to changing one's ideas. One frequently needs to rub out lines that are hidden behind other parts.
>
> (p. 91)

So sketching is one of the main aids to creative thinking in invention and design, which is why we encourage you to use it whenever you can. For visualising certain three-dimensional designs, for example mechanisms involving movement of complex shapes, sketching is not adequate and it may be necessary to make simple physical models, say using cardboard cut-outs and drawing pins for rotation axes. (On Video 3, Section 3, 'The challenge of the portable bike', you will see how both sketches and simple models were used in the conceptual design of the Strida folding bike.)

Sketches, drawings and models are aids to help the inventor or designer *visualise* the idea or object they are try to create. But in addition to this many designers and inventors are able to visualise their ideas 'in their mind's eye' even before sketching or modelling them. This extends from the interior designer who must visualise the results of choosing certain furniture, curtains and wallpaper in a given room to electronic engineers who have to have a clear mental picture of component characteristics so that they can quickly consider the effects of choosing one possible circuit over another. The electrical engineer Nikola Tesla claimed that he successfully visualised and 'ran' models of his alternating current motor in his mind with the result that his first motor worked perfectly. Thring mentions Buckminster Fuller, inventor of the geodesic dome, as a designer with a particularly highly developed three-dimensional visualisation skill gained through practical experience:

> Buckminster Fuller who has produced so many inventions in the field of static structures was brought up on a small island where sailing ships and their highly developed structures of ropes and spars were part of his daily physical education. There is no doubt that his exceptional ability to think in three dimensions and visualise the forces leading to static stability gave him the necessary freedom to imagine entirely new ways of supporting objects in space and to design on the surface of a sphere.
>
> (pp. 89–90)

So, in addition to the ability to switch rapidly between divergent and convergent modes of thinking, the inventor or creative designer should be able to visualise three-dimensional objects both static and moving, to sketch, draw and make models and to have a practical feel for the real world of objects and systems.

5.3 KNOWLEDGE AND CREATIVITY

Most of the intellectual and practical skills outlined above are not a matter of inherent talent, but can be acquired through practice or by the application of techniques. There still remains the question of how much knowledge and experience is required to produce worthwhile inventions and design ideas in a particular field.

There are some who argue that too much knowledge of a particular subject or experience in a particular area inhibits creative thinking because of over-familiarity with existing ideas and the tendency of the mind to become 'set' along conventional lines. Jewkes *et al.* say:

> The essential feature of innovation is that the path to it is not known beforehand. The less, therefore, an inventor is pre-committed in his speculations by training and tradition, the better the chance of his escaping from the grooves of accepted thought. The history of invention provides many examples of the advantages, if not of positive ignorance, at least of a mind not too fully packed with existing knowledge or the records of past failures.
>
> (Jewkes, Sawers and Stillerman, 1969, p. 96)

They provide some evidence for this view by citing several examples of inventions 'where a fresh and untutored mind has suceeded when the experts have failed'. Their examples include Biro, inventor of the ball-point pen who was an artist and journalist; Dunlop, the inventor of the pneumatic tyre, who (you will recall from Section 4) was a vet; and Gillette, inventor of the safety razor, who was a commercial traveller.

On the other hand Jewkes and his colleagues also provide many examples of inventions originating from individuals with a professional scientific or engineering background, although this was not always in the specific field of the invention. For example Lee De Forest, inventor of the three-element radio valve, was a mechanical engineer; while Christopher Cockerell, inventor of the hovercraft, was an electronics engineer turned boat-builder.

It is now generally recognised that, except perhaps in certain less technical areas, knowledge is usually vital to worthwhile invention and creative design. This does not mean that relevant knowledge is restricted to the specific field in which the inventor or designer is working. In general, the broader the knowledge of the inventor or designer the better. Leonardo da Vinci, perhaps the greatest inventive genius of any age had an extraordinarily wide knowledge of existing and past technology as well as of subjects ranging from mathematics, geology and anatomy to music. It is not excessive knowledge that causes someone to think along conventional lines so much as habit and mental inflexibility. This is as true of a beginner in design as of Leonardo. Lawson (1980) notes before student designers can begin to produce creative ideas they need a pool of experience and technical knowledge to draw upon, but at the same time need to retain the flexibility of mind to allow that knowledge to be patterned and combined in new ways. As Michael French points out, mental flexibility is enhanced if the designer's knowledge is retained in terms of general *principles* as well as specific examples and solutions.

> The mind of the designer should be like a rich, open soil, full of accessible resources, of which the chief is an ordered stock of ways and means, illustrated by many examples. The sum total of all such stocks might be called the 'design repertoire'....the repertoire should be organised on the most abstract lines possible, so that any item in it is likely to be recalled in as many contexts as possible: too particular a label may cause it to be overlooked.
>
> (French, 1988, p. 300)

Building up a **design repertoire** is partly a matter of experience, but, as French notes, there are many excellent reference books containing principles and solutions for both general and specialised areas of design.

5.4 ADAPTORS AND INNOVATORS

So far I have discussed the characteristics and abilities of individuals that affect their capacity for creative work, especially in product invention and design. However, this does not necessarily take into account the fact that different people tend to approach similar problems that require creative thinking in different ways.

In the 1960s a British occupational psychologist, Michael Kirton, found that managers tended – to a greater or lesser extent – to tackle similar organisational problems in one of two characteristic ways. Some managers tended to produce ideas and solutions which involved improvements to the existing system: Kirton called this group *adaptors*. Other managers tended to propose ideas and solutions which involved radical changes to the existing order of things: Kirton called this group *innovators*. Both adaptors and innovators could produce creative solutions to a problem, but they differed greatly in the degree of change their solutions required. Adaptors produced generally acceptable ideas for 'doing things better', while innovators proposed ideas that were often rejected because they involved 'doing things differently' (see for example, Kirton, 1980).

So far Kirton's adaptation-innovation theory has been applied mainly to the assessment and training of managers. However, the notion of there being individuals who tend towards being either adaptors or innovators seems to have interesting parallels with the tendency of some designers to tackle problems by proposing incremental improvements to existing products, while others think in terms of radical new concepts. Some designers certainly seem to have a preference for, and are better at, making evolutionary improvements to existing products, while others are better at creating innovative design ideas. Although in this Block I have tended to focus on 'innovator' designers who have produced major inventions and novel designs, it is important not to forget the, often highly creative, contribution of 'adaptor' designers who have been responsible for translating the first prototypes and early innovations into effective and acceptable products.

SAQ 13

What is meant by convergent and divergent thinking? What mental abilities are involved in divergent thinking? Why are both types of thinking needed for creativity in design?

5.5 AUDIO: THINKING STYLES EXERCISES

At this point you may be feeling somewhat downhearted at the many personal characteristics, abilities and skills that appear to be needed to design something new. So in the next 30 minutes to an hour I would like you to try a few exercises which test your ability to see things spatially, to visualise two-dimensional drawings in three dimensions, to think divergently and convergently, and solve problems involving logical and lateral thinking. They are not tests of intelligence, and many do not have a single 'correct' answer.

Don't worry if you find that you can only do a few of these exercises. They test general mental abilities for tackling different kinds of problem, and are not necessarily a guide to an individual's potential for creative design work. What the exercises can do is to provide some indication of what kind of thinker you tend to be: would you be best suited to open-ended problems involving visual skills or tighter problems involving more logical thinking or systematic analysis? Are you an 'innovator' best at invention or conceptual design, or an 'adaptor' who prefers more detailed design work?

Stop reading now (or, if that is not convenient, wait until the end of the section) and find the *Audio 2 Study Guide*. Follow the instructions for undertaking the exercises given in the Study Guide before listening to the discussion on Side 1, Part 2 of the audio-cassette for this Block, which is entitled 'Thinking styles exercises'.

5.6 CREATIVE DESIGN PROBLEMS

You may object that the questions and puzzles you attempted in the *Audio 2 Study Guide* are artificial exercises and not real design problems. So spend 15 minutes considering the following three real industrial problems (adapted from Ledsome, 1987; Walker, Dagger and Roy, 1991), which are in order of increasing difficulty. The first and third problems were originally challenges which appeared in a magazine for innovative designers called *Eureka!* I do not necessarily expect you to come up with solutions even to the first problem, certainly not within a few minutes. But you should be able to see what kind of thinking, knowledge and skills might be required to produce a solution.

EXERCISE A USING BROKEN HACKSAW BLADES

Hacksaw blades (Figure 68), in the hands of both amateurs and professionals, often snap in use. New blades are not cheap and the broken blade often has much life left in it.

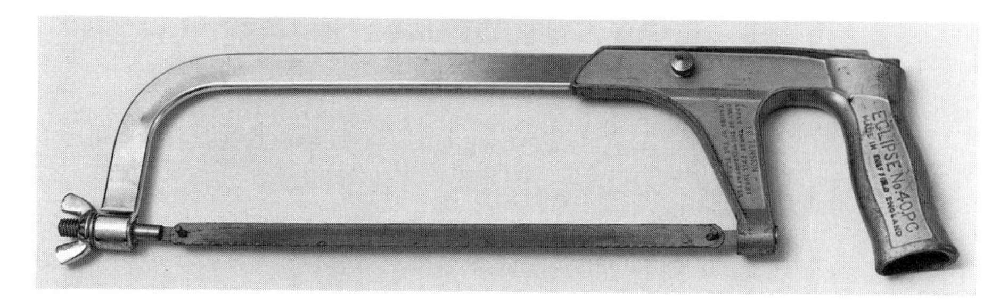

FIGURE 68
HACKSAW

Design a simple gadget to make best use of the broken hacksaw blade, ideally one that would enable the sawing task to be completed.

EXERCISE B HEAT PUMP VALVE

Heat pumps are devices which enable the low temperature heat available in the air, soil, rivers, etc. to be boosted to a useful temperature for heating buildings, swimming pools, etc. They operate by using a compressor to pump fluid through an evaporator, expansion valve and condenser, rather like a refrigerator in reverse (Figure 69).

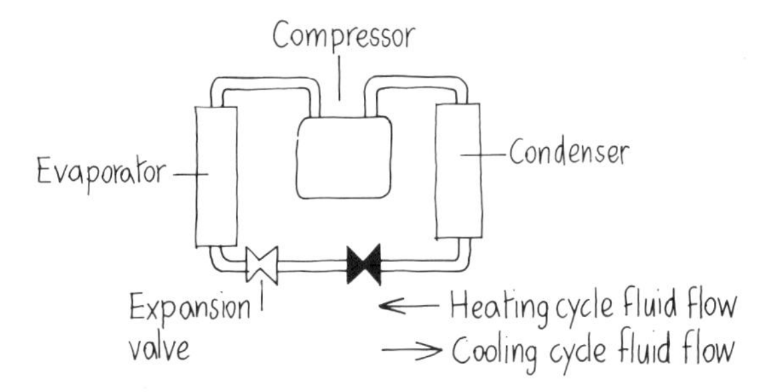

FIGURE 69
HEAT PUMP VALVE PROBLEM

One type of heat pump, which can be used for air-conditioning or space heating, requires twice as much fluid to be pumped when the fluid is flowing in the cooling mode than when the direction is reversed for heating. Suggest a design for a valve that will allow twice as much fluid to be pumped in one direction as the other.

EXERCISE C AUTOMATIC TEA-MAKER

Not all automatic tea-making equipment for domestic use correctly fulfils the long-standing tea-making tradition of warming the pot before brewing starts. This is one reason why automatic tea-makers have a reputation for making poor tea.

Produce a design concept for an automatic machine, suitable for the domestic consumer market, that boils water, introduces to it tea or coffee (in leaf, bag or ground form), prevents further boiling while the liquid brews, and finally, stores the drink in a warmed pot until required.

Stop and attempt the above exercises before looking at the solutions produced by the designers who actually tackled the problems.

Below are some solutions to the above problems which might be regarded as creative, not just because they are novel, but for their simplicity, elegance or economy. As there are usually no correct answers in creative design, you may have thought of something different, or perhaps better.

SOLUTION TO EXERCISE A

See Figure 70.

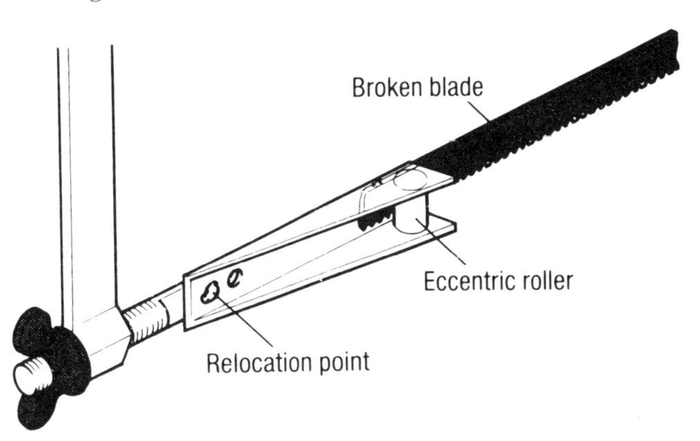

FIGURE 70
'BLADESAVER' DESIGNED AND PATENTED BY MICHAEL BILNEY USES THE SIMPLE PRINCIPLE OF AN ECCENTRIC ROLLER TO GRIP AND MAKE MAXIMUM USE OF THE REMAINING LENGTH OF BROKEN HACKSAW BLADE. IT COULD BE A BETTER WAY OF FITTING THE BLADE IN THE FIRST PLACE. (*SOURCE*: LEDSOME, 1987)

SOLUTION TO EXERCISE B

I could only think of various mechanisms for monitoring the direction of flow coupled to a device which opened and closed a valve to stop part of the flow in the heating cycle. However, the solution produced by one manufacturer (Figure 71) is both simple and elegant. It uses the change in direction of fluid flow to open and close a simple ball valve.

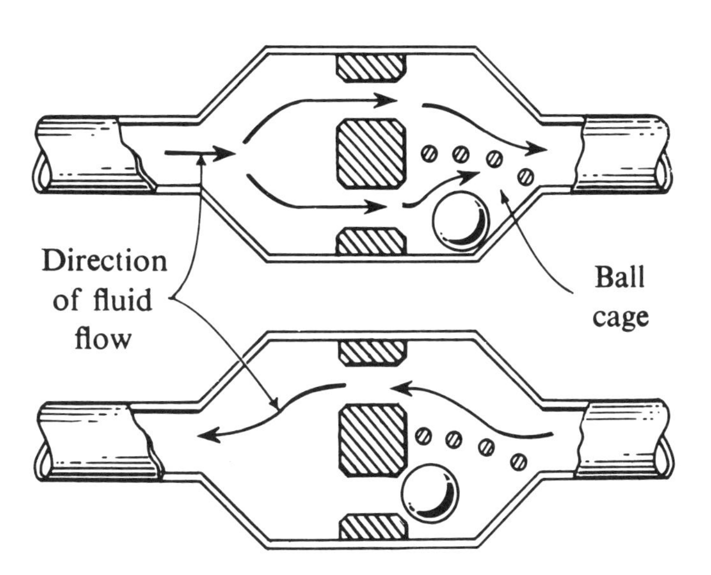

FIGURE 71
ELEGANT SOLUTION TO HEAT PUMP VALVE PROBLEM. THE CHANGE IN DIRECTION OF FLUID FLOW OPENS AND SHUTS THE SIMPLE BALL VALVE (*SOURCE*: MIDDENDORF, 1969)

SOLUTION TO EXERCISE C

Designers at Russell Hobbs used the analogy of a capsizing boat to create a novel solution to the problem.

Their tea-making machine uses a ballasted plastics container with a perforated lid to hold either tea leaves or coffee grounds. This floats with its dry cargo in an electrically heated water jug. When the water boils its surface becomes agitated and so the boat quickly sinks. The ballast used is a magnet and when the submerged vessel reaches the bottom of the jug this triggers a switch to turn off the heating element.

The tea or coffee can be left to brew as long as desired with the heating jug now acting as a tea or coffee pot.

6 CREATIVITY IN PRACTICE: TWENTIETH-CENTURY BICYCLE DESIGN

As you saw in Sections 2 and 4, from the turn of the century to 1960, despite the efforts of many inventors and designers, the bicycle remained fixed on the diamond-frame configuration and underwent gradual refinement of its materials, components and accessories, coupled with improved manufacturing technology.

> Why was the diamond-frame bicycle gradually refined, but hardly changed in basic design?

> The obvious answer is that the bicycle had evolved into an optimum design, and no further change was needed. But this was not the only factor.

Following the First World War, bicycles became the main means of personal transport for those who could not afford motor cycles and cars. The conventional bicycle was satisfactory for this utilitarian purpose, and, apart from the addition of components such as gears and electric lamps, there was no impetus for the industry to innovate in design. There was, on the other hand, a need to make bicycles at prices that ordinary people could afford, and so the larger cycle firms introduced important innovations in *manufacturing* technology, including automatic machinery and production methods developed during the First World War. Thus, as the cycle industry matured, there was a shift from the development of new designs to the efficient production of standard models at low prices – in other words a shift from **product** to **process innovation** (Figure 72). You should recall this shift (which applies to mass-production more than to batch-production industries) from Block 2, Section 6 and from Video 3, Section 1, 'The evolution of the bicycle'. Designing innovative manufacturing systems may involve as much, or more, inventive and creative effort as designing the products they make – as you will see in Block 5. For example, one such process innovation is a flexible manufacturing system that uses computer-aided design and robotics to enable a Japanese bicycle firm to offer affordable custom-built machines to individual customer orders (Figure 73).

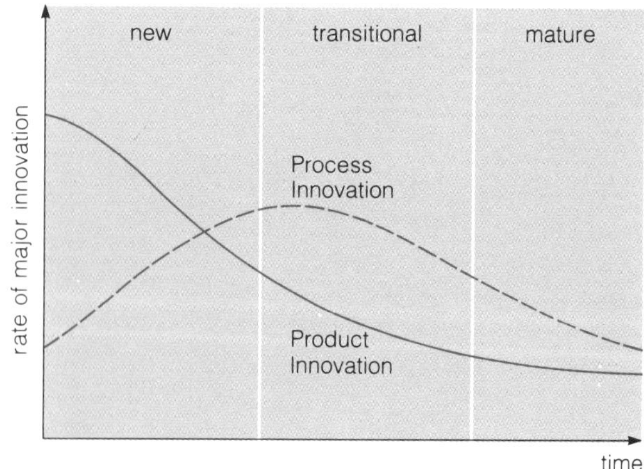

FIGURE 72
SHIFT IN EMPHASIS FROM PRODUCT TO PROCESS INNOVATION AS AN INDUSTRY MATURES

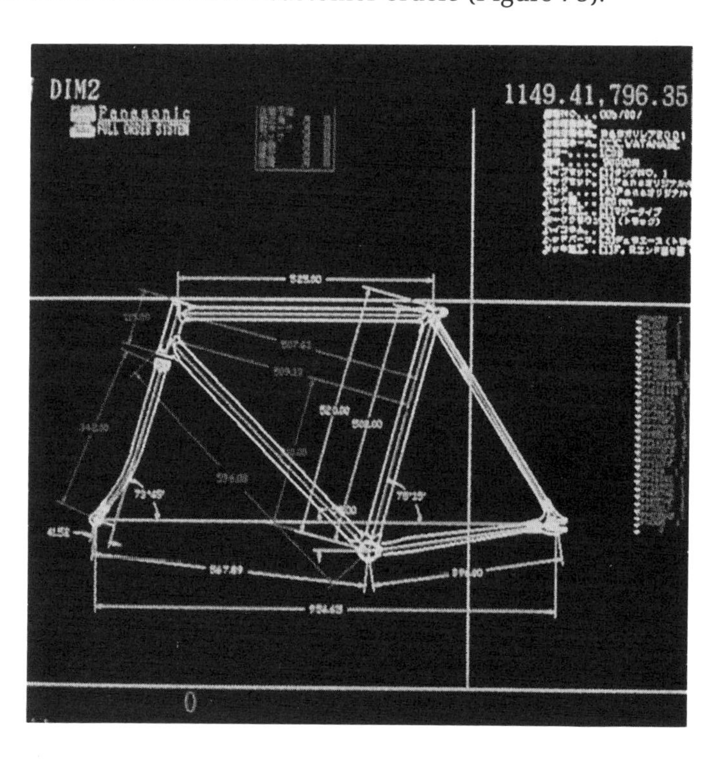

FIGURE 73
COMPUTER-AIDED DESIGN USED TO ENABLE A JAPANESE FLEXIBLE MANUFACTURING SYSTEM TO PRODUCE AFFORDABLE CUSTOM-BUILT BICYCLES TO INDIVIDUAL CUSTOMER'S ORDERS

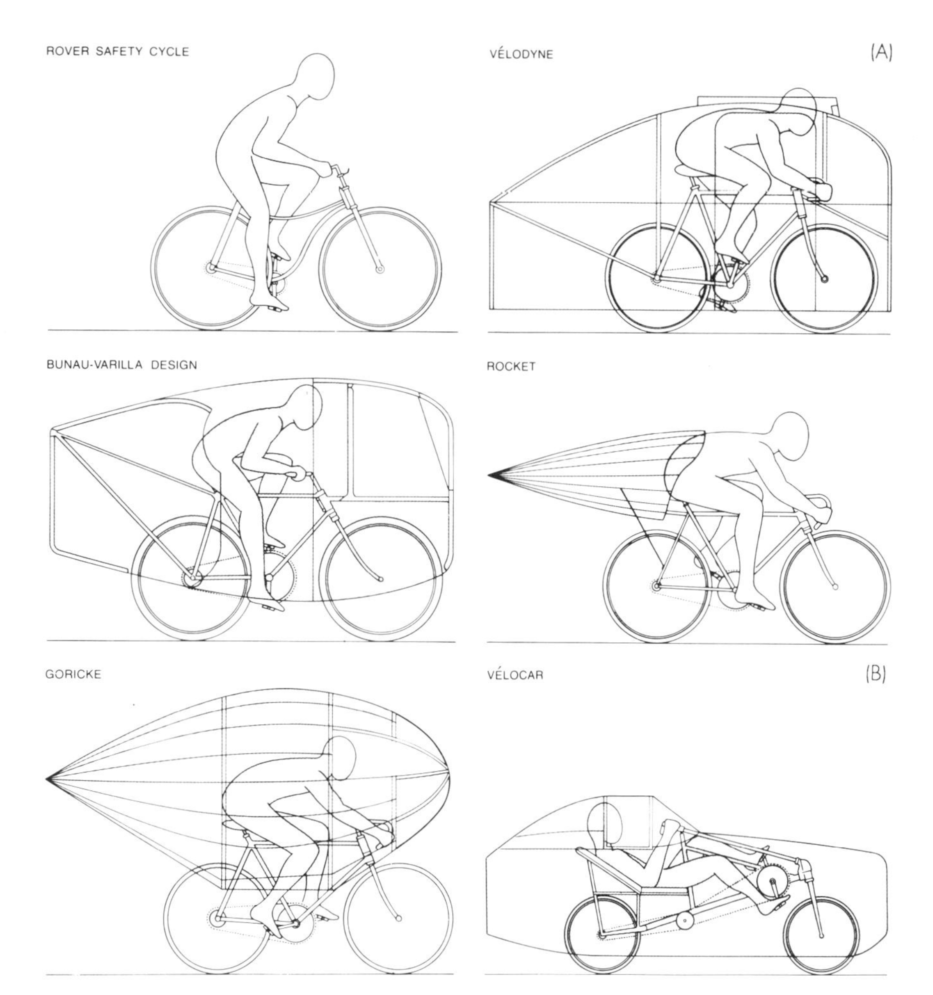

FIGURE 74
AIR RESISTANCE IS THE MAIN FACTOR LIMITING THE SPEED OF HUMAN-POWERED VEHICLES. EARLY ATTEMPTS TO IMPROVE BICYCLE AERODYNAMICS INCLUDE THE BUNAU-VARILLA'S DESIGN OF 1912 AND GORICKE'S OF 1914. KNOWLEDGE OF AERODYNAMICS FROM AIRCRAFT AND CAR DESIGN WAS USED IN THE INVENTION IN FRANCE OF THE VÉLODYNE (A) AND VÉLOCAR (B). SUCH VEHICLES, WHICH BROKE SPEED RECORDS IN THE 1930S, WERE BANNED FROM CYCLE SPORT, AND THIS DELAYED FURTHER DEVELOPMENT OF STREAMLINED HUMAN-POWERED VEHICLES UNTIL THE 1970S. (*SOURCE*: SCIENTIFIC AMERICAN, DEC. 1983)

The other main use of bicycles was for sport. Here there *was* an incentive to innovate to produce lighter and faster machines. This led to the invention of derailleur gears, cable-operated brakes and the adoption of alloy steels and aluminium for frame and component construction. But, unlike cars and aircraft, bicycle design was not seen as an area worthy of serious engineering attention. Even when engineering effort led to inventions such as the aerodynamic *Vélodyne* and *Vélocar* (Figure 74), as you saw on the video, the conservatism of manufacturers, cyclists and the body governing cycle sport meant that innovation was effectively stifled.

After 1960, as you saw in Section 2 of the video, bicycles began once more to exhibit a greater variety of forms, despite the continued dominance of the diamond-frame design. In this section I shall therefore examine some examples of innovations in cycle design that have been introduced since the early 1960s. As before I have had to be very selective in my choice of examples, as there are too many cycle inventions, designs and innovations to include more than a very few. I have concentrated on designs for everyday use, and so have not included any case studies of the various recumbent cycles and streamlined human-powered vehicles (HPVs) that have appeared in the past twenty years. Two examples of the latter were illustrated in Section 2 of the video and some more are shown here in Figures 75 to 77. On the video you can see in action some of the innovative designs of recumbents and HPVs that have been stimulated by the contests held since 1974 by the International Human-Powered Vehicle Association.

FIGURE 75
THE AMERICAN 'AVATAR 2000' WAS THE FIRST OF THE MODERN GENERATION OF COMMERCIALLY PRODUCED RECUMBENT BICYCLES DESIGNED FOR GENERAL ROAD USE. IN BRITAIN IT WAS PRICED AT ABOUT £1200. IN A MODIFIED VERSION WITH A FULL AERODYNAMIC FAIRING THE AVATAR 'BLUEBELL' SET A WORLD BICYCLE SPEED RECORD OF 52 MPH IN 1982

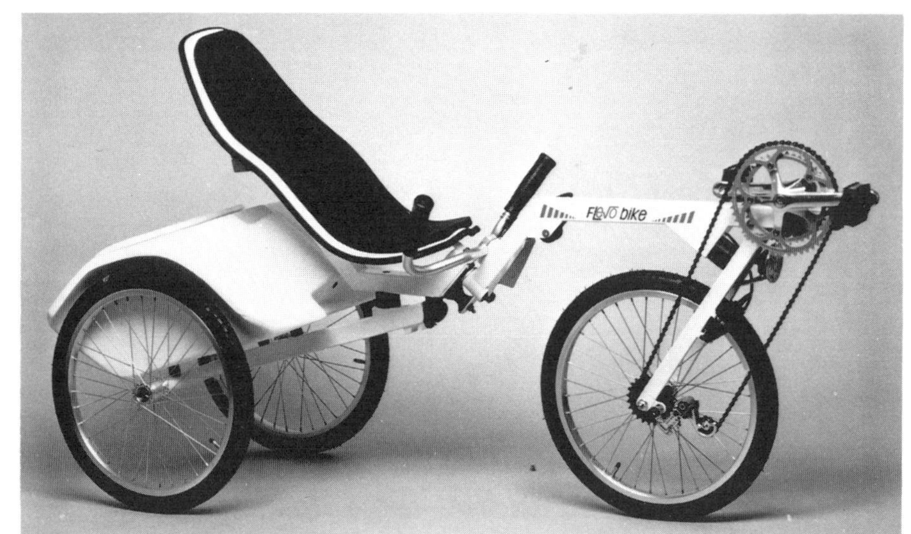

FIGURE 76
THE GERMAN 'FLEVO-BIKE' LAUNCHED IN 1990 IS A NEW DESIGN OF RECUMBENT CYCLE SUITABLE FOR ROAD USE. IT IS AVAILABLE IN BICYCLE AND TRICYCLE VERSIONS

(A)

(B)

FIGURE 77
'THE BEAN', A BRITISH-DESIGNED TWO-WHEELED HUMAN-POWERED VEHICLE:

(A) UNDERGOING WIND TUNNEL TESTS

(B) WITH ONE HALF OF THE SHELL REMOVED

THE VEHICLE SET A WORLD RECORD IN 1990 OF NEARLY 47 MILES TRAVELLED IN ONE HOUR

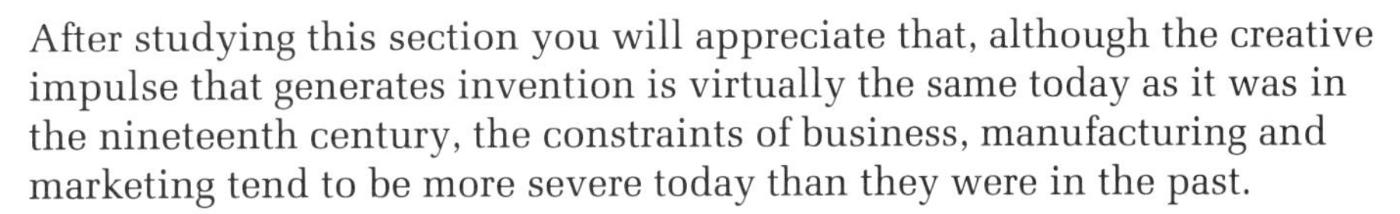
After studying this section you will appreciate that, although the creative impulse that generates invention is virtually the same today as it was in the nineteenth century, the constraints of business, manufacturing and marketing tend to be more severe today than they were in the past.

Three of the examples in this section are presented partly on video. If you have not done so already, you should now view the third sequence of Video 3, Section 1, 'The evolution of the bicycle' which contains material on the Moulton bicycle and the mountain bike. Use what you have learned to answer SAQs 14 and 15 in this section of the Block text and the SAQs in Section 1 of the *Video 3 Study Guide*.

6.1 THE MOULTON BICYCLE

ORIGINS OF THE SMALL-WHEEL BICYCLE

Early cycles had large wheels. The driving wheel of an Ordinary bicycle had a diameter of four to five feet, and even the Rover 'dwarf' safety bicycle of 1884 was equipped with three-foot diameter wheels. By the end of the nineteenth century, however, following the introduction of the pneumatic tyre, the diameter of wheels had been standardised at between 26 and 28 inches, a size that gave a satisfactory ride on ordinary roads and fitted nicely into the adult diamond-frame configuration.

While the classic design has proved itself over the years to be excellently fitted to its purpose, it does have some drawbacks.

FIGURE 78
ALEX MOULTON RIDING HIS SMALL-WHEEL BICYCLE

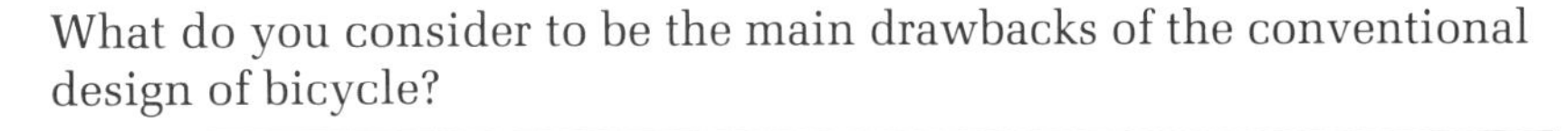
What do you consider to be the main drawbacks of the conventional design of bicycle?

It is rather large for convenient transport and storage.

Different frames are required for people of different heights, and for men and women.

It has a limited load-carrying capacity.

Mounting and dismounting a diamond-frame bicycle can be awkward.

A variety of small-wheel bicycles, including folding types, have been developed in an attempt to overcome these drawbacks.

The idea of a small-wheel bicycle is not new; a prototype small-wheeler with a cross-frame was built in Paris as early as 1868. However, it was not until 1959 that a serious attempt was made to design and introduce a small-wheel bicycle. This was the bicycle invented by Alex Moulton, with sixteen-inch diameter wheels and a rubber suspension system, which was launched on to the British market in 1962 (Figure 78).

The lessons to be learnt from the manufacture and sale of the Moulton bicycle are similar to those of the Dursley Pedersen discussed in Section 4. I shall not, therefore, dwell on these aspects. Moulton's design achieved some commercial success for a few years before being bought up by TI-Raleigh. It was eventually phased out of production in favour of simpler, unsprung small-wheel designs. The significance of the Moulton bicycle was not its invention or its unique rubber suspension system, but the new image it gave to cycling and the stimulus it gave to cycle design. This, coupled with modern marketing techniques, helped in the 1960s and 1970s to revive an industry that, in Britain at least, had been in serious decline for over a decade.

Alex Moulton had intended to design a technically superior, all-purpose bicycle that he believed could rival the classic diamond-frame design. What he did not foresee were the business, manufacturing and market constraints that would adapt his invention into the technically less sophisticated, but commercially very successful, small-wheel 'shopper' bicycles that appeared from the mid-1960s onwards, plus the parallel development of Chopper-type 'fun' bikes for children.

INVENTION AND DESIGN OF THE MOULTON BICYCLE

It is worth examining the invention, design and development of the Moulton bicycle, as this process has been unusually well documented.

As with most inventions, the context is significant. Alex Moulton trained as a mechanical engineer at Cambridge University, and worked in the aircraft industry before joining the family rubber manufacturing firm of Spencer Moulton Ltd. The firm made rubber springs, and Moulton had the idea of developing rubber suspension springs for road vehicles. In 1956 Moulton formed his own firm for research and development, Moulton Developments Ltd., in order to concentrate on the development of vehicle suspension systems under the sponsorship of Dunlop and the British Motor Corporation. In collaboration with Alec Issigonis, designer of the Morris Minor and the BMC Mini, Moulton patented and develope d the rubber cone suspension system first used in 1959 on the Mini, and he subsequently developed the Hydrolastic suspension system used from 1962 onwards on the BMC 1100/1300 series cars.

In 1956, at the same time that he was developing the Mini suspension system, Moulton's boyhood interest in cycling was revived as a result of the petrol shortages caused by the Suez crisis.

Several accounts of how Moulton came to develop his small-wheel bicycle have been published. The one I should like you to read is extracted from a lecture on 'Innovation' that Moulton gave to the Royal Society of Arts in 1979. After the extracts I have posed two questions. You should attempt to answer these before looking at my answers and comments.

FIGURE 79
SCALE MODELS MADE PRIOR TO CONSTRUCTION OF THE FIRST PROTOTYPE MOULTON BICYCLE

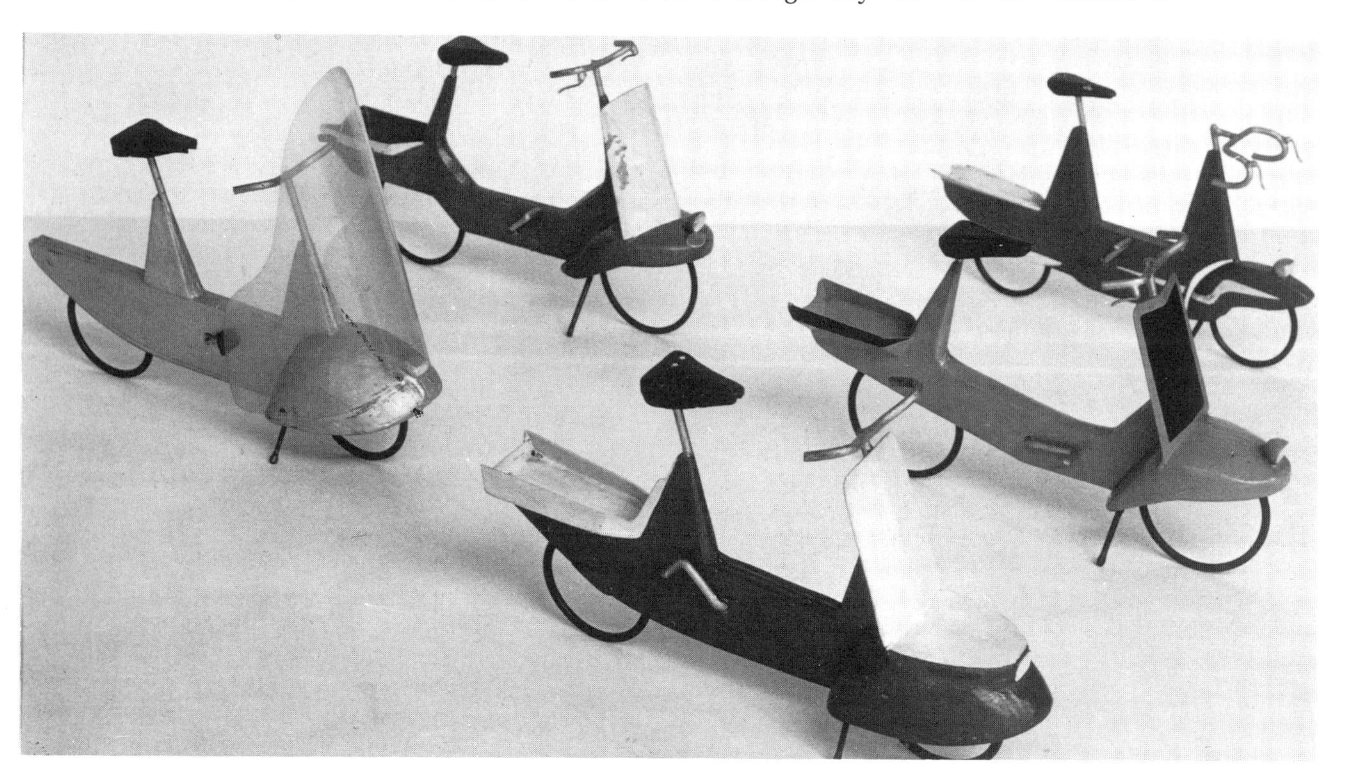

FIGURE 80
FIRST PROTOTYPE MOULTON BICYCLE WITH LIGHT ALLOY SHEET, MONOCOQUE CONSTRUCTION

[...] it was the Suez crisis in 1956 which spawned the Mini as the economical car, and got me interested again in cycling. I had had an early fascination and fondness for cycling as a boy, [...] but it was not until I was in my mid-thirties that I took a serious interest in the machine itself, by using one – and a good lightweight machine it was – under the spur of the petrol shortage. I pondered on what it comprised and how it had evolved to what it was. Having at that time established the laboratory for the car suspension development I had some facilities for making design studies leading to actual hardware prototypes.

The question of the minimum 'kit of tools' for any innovatory achievement is an important one. 'Good ideas' are two a penny but it is hammering them into the harsh reality of a working device that is the difficult thing, and this demands some sort of making facility. An engineer can (unfortunately) achieve no realisation without the facility of making.

I resolved therefore in 1959 to 'have a go' at improving the design of the bicycle – to make it more pleasing to use and possess. The first thing, as always, was to make a study of the fundamentals. I questioned the riding position of the classic machine and acquired a pre-war Grubb to carry out tests of the alternative reclining [recumbent] position. Although good for a short period of activity it was tiring for the longer exercises. I confirmed, therefore, the virtues of the 'classic' position.
[...]
I questioned also the size of the wheel: remembering that in all other vehicles the reduction in size of wheel has been in the direction of design evolution because so many advantages generally speaking flow from it. We were just entering the start of the Mini era with its small wheels and my rubber suspension, and it was not difficult for me to get designed and made by Dunlop a 14 in and 16 in wheel and tyre with a 'proper' one-and-three-eighths section. It would have

been a false experiment to have attempted to use the 'juvenile' type of tyre section and construction. This is an example of carrying out the 'key experiment' properly: and here is the first full-size prototype [Figure 80]. You will see the 'open' frame, which was the innovation of the unisex principle, and the carrying capacity front and rear set low down due to the small wheels. I had feared of course the effect of the reduction in 'flywheel effect' in reducing the size of the wheel. In all engineering innovations, which later we take for granted, the originator will have gone through agonies of doubts. Certainly I did over this. In fact the balancing function of a bicycle is highly complex: and there are advantages as well as disadvantages of small wheel size, and in the event the behaviour was totally accepted. What was apparent was the necessity of wheel suspension to make the small diameter narrow section tyres acceptable from shock transmissions. But in providing suspension the consequence was a total gain in riding comfort. Now on the rare occasion when I ride a classic machine I wonder how anybody can endure the vibration, but the human being is very tolerant!

I was quite wrong in my original choice of the structure. Coming as I had from aeronautical origins by way of contact with the motor car it was natural for me to choose a light alloy sheet metal construction. It was immediately apparent in the reality of the testing that the high concentration of loads in a bicycle demanded a 'boney' structure. Moreover, the 'boxy' sheet metal frame introduced to the bicycle the outrage of road excited noise.

Then quickly followed a period of development of the optimised construction [Figures 81 and 82]. When a problem is accurately enough posed then finding the solution can be quick but it is important to scan all possible routes to ensure optimisation.

(Moulton, 1979, pp. 40–41)

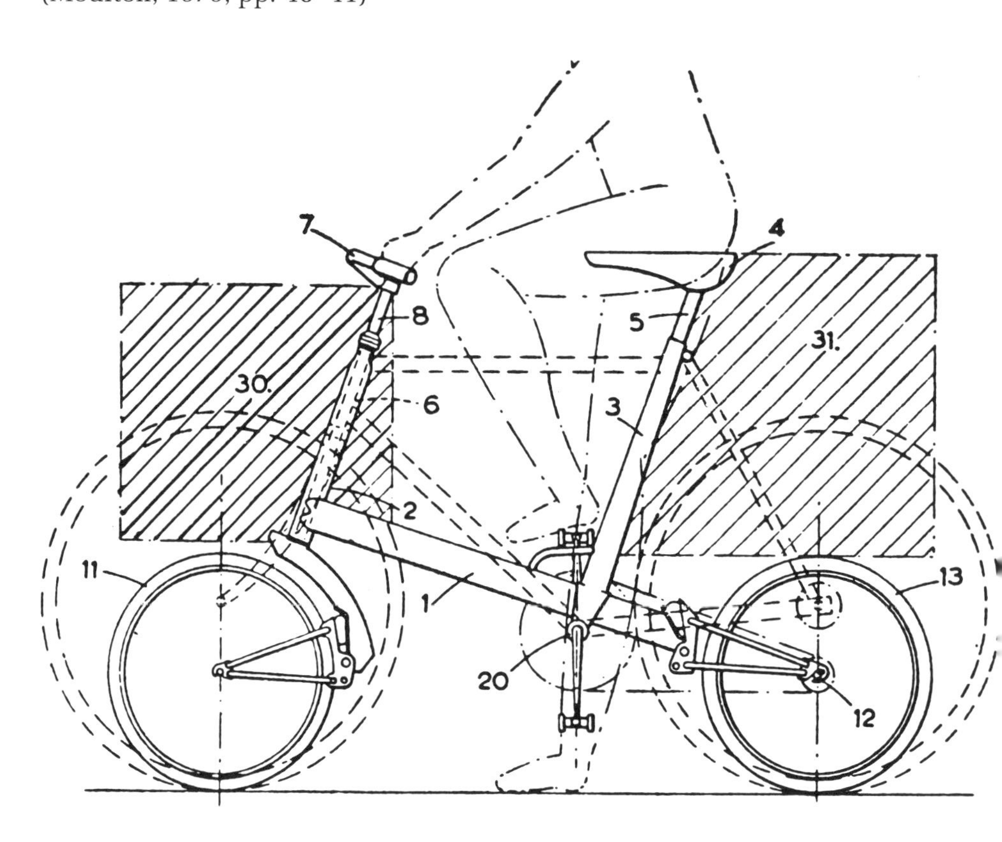

FIGURE 81
ORIGINAL PATENT DRAWING OF THE MOULTON BICYCLE

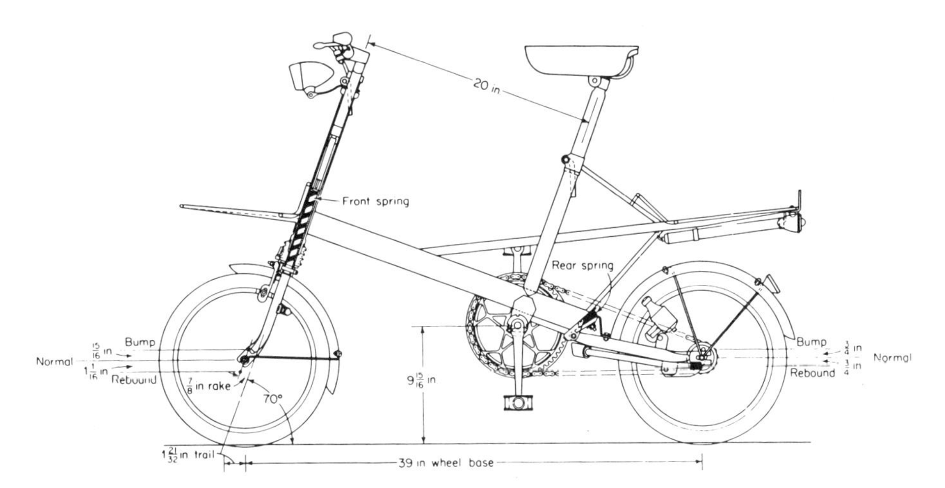

FIGURE 82
GENERAL ARRANGEMENT ENGINEERING DRAWING OF THE SERIES ONE MOULTON BICYCLE

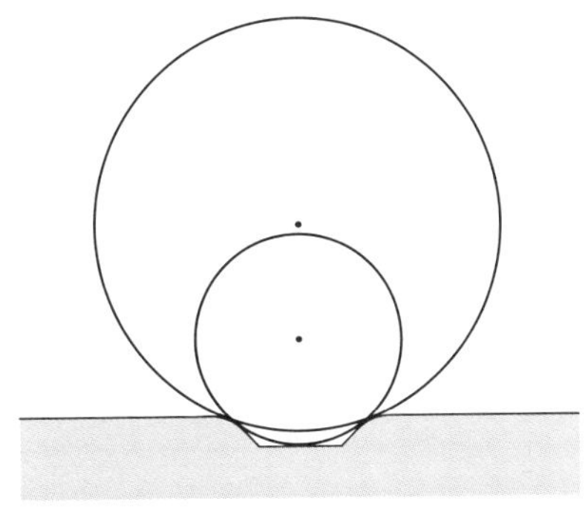

FIGURE 83
A SMALL WHEEL FALLS FURTHER INTO A HOLE THAN A LARGE ONE

Apart from the small wheels and the F-shaped cross-frame, the most unusual feature of the Moulton bicycle was the rubber suspension system. This was needed becausc small wheels with high pressure tyres are less able to absorb road shocks and will fall deeper into holes than would a large wheel (Figure 83). Given Moulton's background in the rubber industry and his experience with rubber vehicle suspensions, it is perhaps not surprising that his bicycle should have ended up with rubber springing. This suspension system went through several stages of development before the final version emerged. The final design involved a telescopic coil-spring and rubber front suspension (Figure 84) and a trailing-link rear suspension also sprung by rubber. This shows that it is often necessary for a designer to produce and test *several* concepts in the creation of an innovative product.

On the basis of your reading of the Block so far, what do you consider to be the most significant points about the creative process that led to the invention and design of the Moulton bicycle? How does it compare with the process involved in the conception and design of the Rover Safety, the pneumatic tyre and the Dursley Pedersen?

---

1. As in the cases of the Rover Safety, etc. there was an individual who was dissatisfied with existing designs and was determined to improve upon them ('constructive discontent').
2. The designer concerned was prepared to question the fundamentals of the existing configuration (the 'inventor's eye').
3. There was a 'primary generator' for the invention. Moulton considered that in all other vehicles wheel sizes had reduced with time. An experiment confirmed his intuition that small wheels with high pressure tyres would roll just as easily as the conventional type.
4. The previous background and experience of the designer strongly influenced his ideas. Moulton's choice of rubber suspension is one example. Another is his initial idea of using alloy sheet construction. By building a prototype Moulton soon discovered his mistake in the latter idea and for the second prototype he made a 'jerk change' to a tubular F frame configuration which led to the final design.

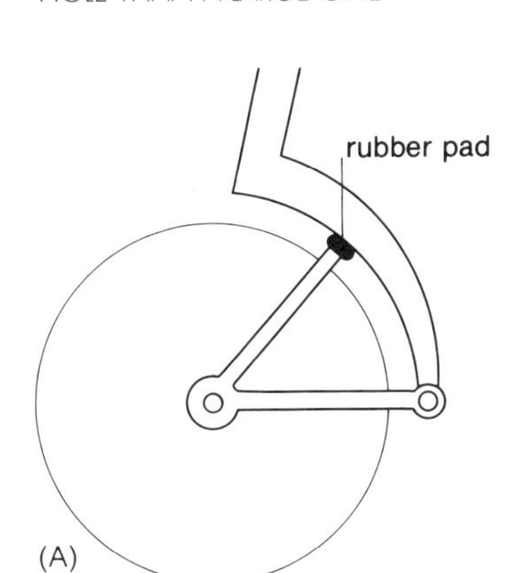

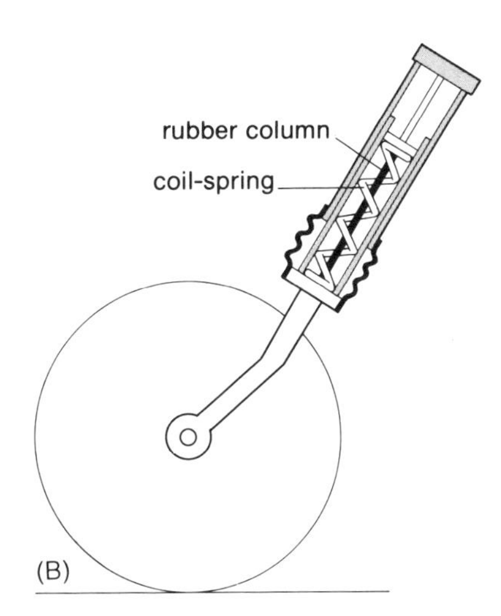

FIGURE 84
TWO STAGES IN THE EVOLUTION OF THE MOULTON FRONT SUSPENSION SYSTEM

However, one thing seems to distinguish Moulton, as a qualified engineer, from his historical predecessors in the field of bicycle design. That is his ideas were based not only on experience and intuition, but on a careful study of the problem to be solved followed by generation and engineering analysis of possible solution concepts and thorough testing

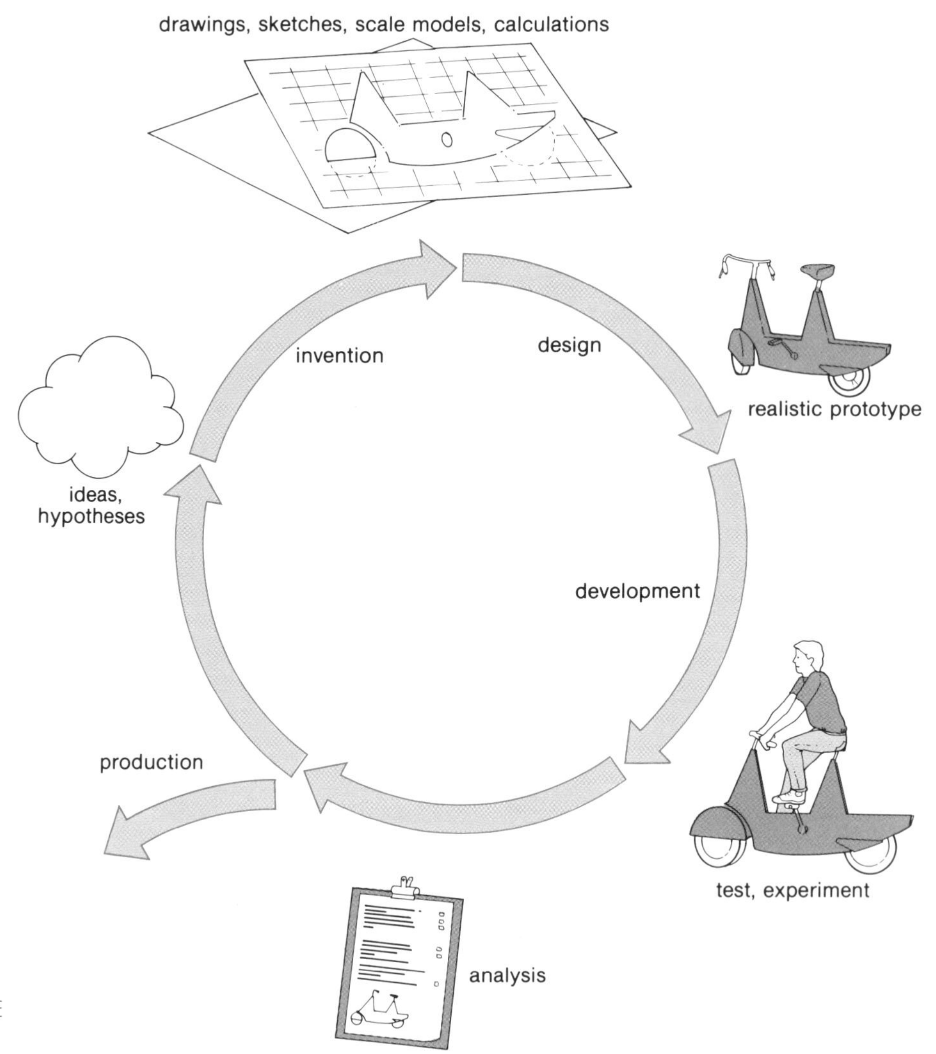

FIGURE 85
MOULTON'S PROCESS FOR INNOVATIVE DESIGN

of a series of prototypes. Moulton's process of innovative design may be represented as shown in Figure 85. The process is repeated as many times as is necessary to optimise an 'immaculate' production prototype. In all, Moulton made and tested fifteen prototypes of his bicycle before the launch of the production model. In his lecture Moulton notes the need for divergent thinking, 'it is important to scan all possible routes [to a design problem] to ensure optimisation'. Elsewhere he has commented on the even greater need to analyse and test ones's divergent ideas:

> Ideas and calculations must be translated into drawings and sketches [...] drawings must be made into hardware as soon as possible, so that reality can be tested and analysed. This is the most important part of the development cycle.
>
> (Whitfield, 1975, p. 175)

And in another lecture he firmly aligns himself with those who stress the need for knowledge and experience in innovative design:

> What differentiates the designer, who successfully innovates, from the crackpot inventor is the depth of study. Certainly I have made [...] dozens of 'inventions' leading to patents; but they all arise from a revelation emanating from observing and studying in a particular field; never from a random idea occurring in a random field.
>
> (Moulton, 1966, p. 1)

INNOVATION AND IMPACT OF THE MOULTON BICYCLE

In common with many creative designers, Moulton had difficulties in getting his ideas accepted. Moulton continued his 1979 Royal Society of Arts lecture by explaining why his invention would probably never have become an innovation had he not possessed the money and determination to manufacture and market the small-wheel bicycle on his own account.

> It was always my intention to produce the design of a new and improved bicycle [...] but not to set up as a manufacturer myself. Little did I realise what was to lie ahead of me! I had long and encouraging discussions with the then chairman of a great bicycle maker, and we got close to a licence agreement. However, it never matured for various factors, which I was to learn are common in business life, such as the consequence of take-over, changes in top management and so forth. But the underlying cause was the 'NIH' [not invented here] factor which was the first time I had come across it.
> [...]
> In my case it was exacerbated because I was offering not only a new design of bicycle, which was full size and yet smaller than the classic, but it was submitted in the form of professionally made and finished prototype machines which could be ridden and put through any test. I well remember, after one of my many meetings with the company, finding one of the officials – a big man – scrubbing the bike *sideways* along the passage with an expression of such intense hatred towards it. He was actually wanting to break it. (In fact it was very strong, so he couldn't!) He must have regarded it as a threat to the stability which he found in the finality of the familiar and fixed design of the classic bicycle.
> [...]
> It was dawning on me, while I was responding to the real or imagined criticism of the design by these experts, with further prototypes, that I had better start thinking about making and marketing the thing myself, such was my personal conviction of its merit and potential. I do urge others who are convinced within themselves of the merit of their innovations to 'have a go' themselves – starting in the smallest way if need be.
>
> Before taking the step of risking my own money, I decided to do another test. Whereas I do not subscribe to the view that Market Research is bunkum, nevertheless in the case of an innovation of design then one must go over the heads of the salesmen to the potential buyer with an actual product for him or her to see and try. This I did by arranging with a market survey firm to ask the key questions to customers entering a bicycle shop chosen to be as far away from the bicycle industry as possible. [...] The result of this tiny sample survey showed an overwhelming positive reaction, quite opposite to that of the experts of the Firms and the Trade.
>
> I decided therefore to go ahead on my own to make and market the bicycle to be called 'Moulton' at a factory which I started in the grounds at home.
>
> (Moulton, 1979, pp. 41–42)

On the evidence of Moulton's article, what personal qualities does an inventor–designer need if he or she is to innovate?

---

1. Interest in and enthusiasm for the problem to be tackled.
2. Theoretical understanding and practical skills needed to develop an idea to the working prototype stage (and the workshop facilities to do this).
3. Sufficient confidence in one's solution to overcome the obstacles to innovation.

As noted earlier TI-Raleigh, then Britain's largest cycle manufacturer, initially declined to produce Moulton's bicycle. The firm was therefore surprised to discover that, following a successful launch in 1962, the Moulton bicycle sold very well. Riders on Moulton bicycles competed successfully against riders on conventional machines in long-distance time trials and also in track races.

These sporting successes naturally helped to promote the Moulton. More important, however, was the fact that the Moulton was visually appealing and became highly fashionable: 'a mini-bike to go with mini-skirts and mini-cars; all part of the 'Swinging 60's' (Hadland, 1982, p. 34). Many famous people were photographed riding Moulton bicycles and this all served to start a revival in the retail cycle trade in the mid-1960s, in an industry that had fallen into a severe decline.

FIGURE 86
THE RSW16 BICYCLE INTRODUCED BY RALEIGH IN 1965 TO COMPETE WITH THE MOULTON

SUCCESSORS TO THE MOULTON BICYCLE

Raleigh meanwhile had commissioned its own market research. This confirmed that conventional bicycles and cycling had an unfashionable, 'cloth-cap' image; there was a latent demand for bicycles, among women and children especially, but what was needed to revive the market was a 'fun means of conveyance'. Raleigh's design department was thus given a marketing specification for a new kind of cycle. 'The bike had to be unisex; it had to cover the range of adults and juveniles; it had to be different in image; suited to traffic conditions and had to reflect the fact that there was more money to spend, and on children.' (Mansell, 1973, p. 88).

Moulton held the patents on the suspension system that permitted the use of narrow-section, high-pressure tyres on the fashionable small-wheel design. So Raleigh's chief designer Alan Oakley conceived a small-wheel unisex bicycle, similar to designs that had appeared on the Continent, that used another method for overcoming the problem of road shocks. This bicycle was named the RSW16 (Raleigh small wheels 16 inch) in order to identify it firmly with Raleigh rather than with Moulton (Figure 86).

Look at Figure 86. What approach do you think was used to conceive the RSW16 design? What was the method used to absorb road shocks on the RSW16?

The RSW16 was essentially an adaptation of existing cross-frame small-wheelers, but with two-inch section 'balloon' tyres inflated to a relatively low pressure (35 psi) to absorb road shocks.

FIGURE 87
SHOPPER BICYCLE WITH 20 INCH WHEELS, SEMI-BALLOON TYRES AND A UNISEX CROSS FRAME BECAME THE DOMINANT SMALL-WHEEL DESIGN, DESPITE ITS TECHNICAL INFERIORITY TO THE MOULTON

Although much more sluggish in performance than the Moulton, because of the higher rolling resistance of the balloon tyres, the RSW16 was cheaper and incorporated several novel features, including a hub gear operated by a twist-grip control and integral dynamo lighting. It sold well, and provided the Moulton with heavy competition. Further competition came from other manufacturers, who by the mid-1960s had introduced rival small-wheelers using a unisex frame and 20-inch diameter wheels, a design compromise between the 16-inch and 27-inch wheel sizes (Figure 87). In 1967 Raleigh introduced its own 20-inch wheel 'shopper' bicycle, aimed mainly at women.

Moulton attempted to counter the increasing competition by rationalising his range of bicycles and introducing a 14-inch wheel design for children. But the scale and determination of his main competitor's marketing operation forced Moulton to sell his cycle company to TI-Raleigh in 1967.

Raleigh retained Moulton as a design consultant, and continued to manufacture Moulton bicycles until 1974, managing to sell over 250 000 Moultons worldwide. But by then, the relatively heavy materials and semi-balloon tyres used on rival designs had led to the image of the small wheel bicycle being inherently harder to ride than the conventional type: an image problem that still dogs the Moulton today.

The decline of the Moulton under Raleigh control came earlier, in 1970, when the company decided to launch simultaneously three new models aimed at different parts of the small-wheel bicycle market: the RSW Mark 3 (marketed as 'the Dolly One'), the Moulton Mark 3 ('the Smooth One'), and the Chopper ('the Hot One') (Hadland, 1982, p. 52). The Chopper (Figure 88) turned out to be such a commercial success that it eclipsed and eventually displaced the other two models.

The success of the Chopper led Raleigh to develop a whole range of childrens' fun bicycles. However, by the mid-1980s these were in turn displaced by the arrival from America of BMX (bicycle moto cross) bikes. BMX machines, which first appeared in the early 1970s, had strengthened and padded frames and strong wheels with knobbly tyres designed for racing over rough ground or for performing daring stunts (Figure 89). In practice most of the bikes sold to young riders during the BMX craze were used for ordinary cycling. Like the fun bicycles before them, BMX machines were sold on their visual appeal and fashionable image (i.e. psychological design factors) as much as, and often in place of, their technical performance and fitness for purpose (i.e. practical design factors). In this sense bicycles have much in common with many other products for the consumer market, as was discussed in Blocks 1 and 2.

FIGURE 88
RALEIGH'S 'CHOPPER' FUN BICYCLE FOR CHILDREN WAS A HUGE COMMERCIAL SUCCESS ALTHOUGH IT WAS CRITICISED ON SAFETY AND OTHER GROUNDS

FIGURE 89
BMX BICYCLES DESIGNED FOR PERFORMING AERIAL STUNTS

THE MOULTON AM ADVANCED ENGINEERING BICYCLES

After Raleigh stopped production of his original small-wheel bicycles Alex Moulton started work on further improving his design, and in 1977 he decided to 'go it alone' once again with a new machine. The aim was to produce a design which was as light in weight as the best general-purpose classic bicycle while retaining the advantages over the classic design provided by small wheels with high pressure tyres and suspension. As before this involved Moulton in creating and sketching several design concepts, analysing them and then making and testing many prototypes of both frame and suspension system.

In evolving the frame Moulton started with a Y-shaped configuration, then, because this was insufficiently light, tried large diameter thin-walled tubes before making the innovative jump to an X-frame (cruciform) configuration. This concept was developed into a multi-tubular space frame design which, as well as being light and strong, is very stiff both laterally and torsionally.

The 'advanced engineering' Moulton AM series was launched in 1983, aimed at the discerning cyclist at the top end of the market. Since then several different models have been designed, including the Moulton Speed for road racing and an all-terrain version suited to off-road riding. The Moulton All-Purpose Bicycle, a model designed to be made in larger volumes at lower cost by another cycle manufacturer was launched in 1992. Figure 90 shows one of the range of Moulton AM bicycles, and you can see Alex Moulton with his new design on Video 3, Section 1, 'The evolution of the bicycle'.

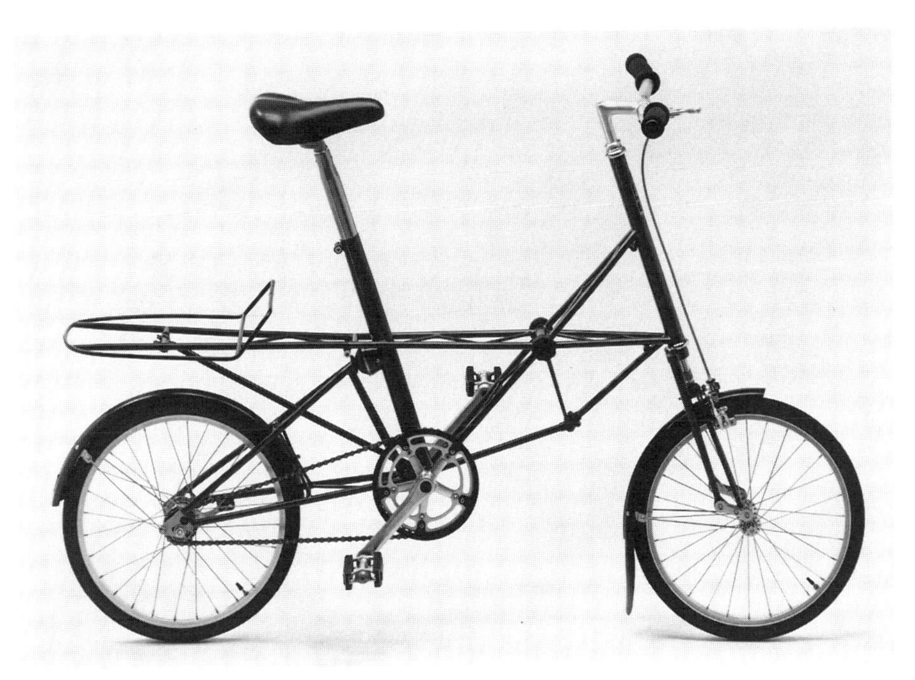

(A)

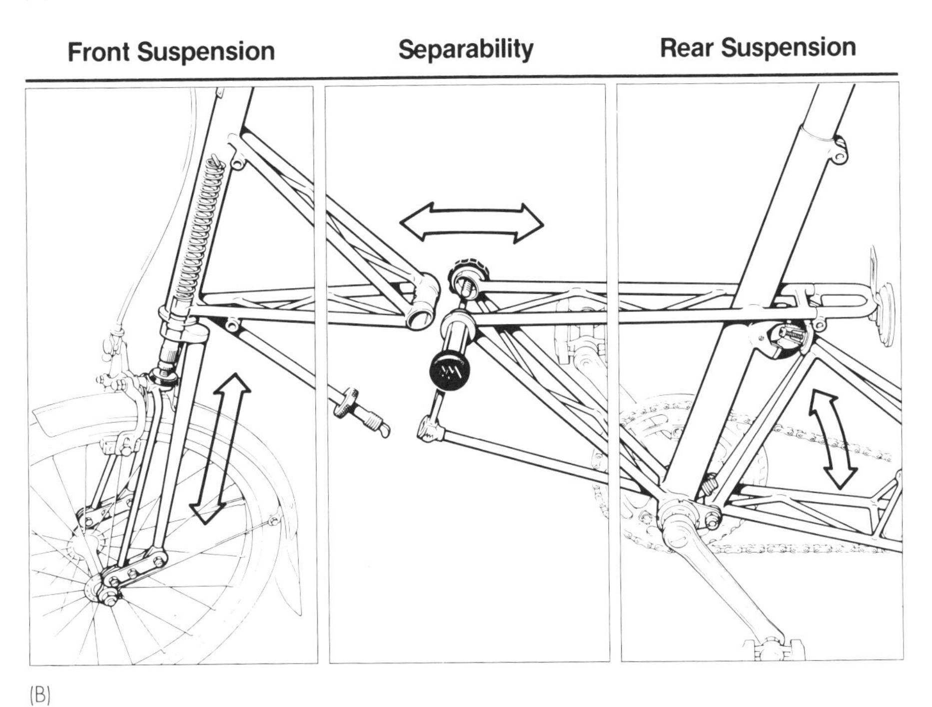

(B)

FIGURE 90

(A) MOULTON AM2 BICYCLE WITH 17-INCH DIAMETER WHEELS. THE AM SERIES IS AIMED AT THE DISCERNING CYCLIST WITH MODELS PRICED (IN 1991) AT £900 FOR THE AM2 TO £2500 FOR THE STAINLESS STEEL FRAME GT

(B) THE NEW FRONT AND REAR SUSPENSION AND LIGHTWEIGHT MULTI-TUBULAR SPACE FRAME THAT ALLOWS SEPARATION OF THE MACHINE INTO TWO PARTS AT A CENTRAL KINGPIN FOR EASY STOWAGE

6.2 MOUNTAIN BICYCLES

Another major cycling innovation in which fashion became as important as function in design is the mountain or 'all terrain' bike. Mountain bikes had their origins in the heavy-duty machines with balloon tyres built by a few north Californian enthusiasts in the late 1970s for downhill racing and off-road riding. The designers of these early so-called 'clunkers' created their machines in the typical manner of enthusiasts by experimenting with different combinations of available components and modifying these by trial and error to improve performance. In 1981 the first purpose-designed mountain bikes began to appear on the mass market in the U.S.A. and reached the U.K. in 1984. By 1990 mountain bikes accounted for over half of all new bicycles sold in developed countries, and had produced a major revival in the fortunes of the cycle industry.

As with BMX, most of these machines hardly ever leave the ordinary road. The main appeal of the mountain bike is its fashionable image, which is promoted with the use of bright colours, special clothing and heavy advertising (Figure 91). As is shown on the video, the mountain bike has been more successful than any design this century in changing the 'cloth cap and cycle clips' image of cycling into one associated with freedom, fitness and affluence.

The popularity of mountain bikes is not of course just due to their psychological design factors. In terms of functional design their toughness makes mountain bikes better suited to many urban and recreational uses than conventional machines and, if desired, they can be ridden off-road too. Mountain bikes have also stimulated important innovations in components and materials. In particular Japanese manufacturers were quick to spot the commercial potential of these new designs and developed suitable new components, such as powerful brakes and derailleur gears that could be quickly and easily changed without the rider losing power.

Raleigh: Nothing comes close.

(A)

(B)

FIGURE 91

ADVERTISEMENT FOR A MOUNTAIN OR ALL TERRAIN BIKE (ATB). ALTHOUGH ORIGINALLY CONCEIVED FOR DOWNHILL RACING AND OFF-ROAD RIDING, MOUNTAIN BIKES ARE BEING MARKETED AS 'STREET FASHION' (RATHER LIKE TRAINING SHOES)

(B) RALEIGH 'DYNA-TECH' MOUNTAIN BIKE WITH BONDED ALLOY STEEL FRAME AND ALUMINIUM LUGS

Despite the dominance of major manufacturers in the mass market, mountain bike enthusiasts continue to design by trial and error. As one of the original inventors of the mountain bike notes:

> Bicycle design is as much voodoo as it is science ... Mountain bikes ... have evolved not so much on the drawing board as in practice. Even the expert builder can only generalise about a bike's performance during the design stage; the proof of whatever theory is being applied comes when he swings the leg over for the first time ... Since mountain bikes are used for such a wide range of activities ... the subject lends itself to endless modification and tinkering.
>
> (Kelly and Crane, 1988, p. 90)

The development of the mountain bike is one of the topics in the third sequence of Video 3, Section 1, 'The evolution of the bicycle'. You should have viewed it by now.

SAQ 14

Why did the small-wheel bicycle conceived by Moulton find its major commercial success in the form of the 'shopper' bicycle and the 'Chopper-type' children's bicycle?

SAQ 15

What are the differences in the approach to design of Alex Moulton and the individuals who created the mountain bike? For what types of design are these approaches best suited?

6.3 VIDEO: THE CHALLENGE OF THE PORTABLE BIKE

For this section you will be using Section 3 of the video entitled 'The challenge of the portable bike'. It shows how designers have responded to the problem of designing a portable human-powered transport device that, unlike a conventional bicycle, is light enough to be easily carried and small enough to be used in conjunction with other means of transport.

One solution to this problem is of course a bicycle which folds, and the video starts with examples of portable bicycle designs based on different objectives and mechanisms (Figure 92). The video then looks in detail at the conception and development of one particular folding bicycle called the 'Strida' which was designed by Mark Sanders (Figure 93). The aim is not to assess the quality of the Strida as a folding bicycle, but to show how Sanders came up with his basic concept of a bike that would fold into a form like a walking stick with wheels, how he created a design that would fit that concept and developed it into a marketable product. The video clearly demonstrates the value of sketching and drawing for design thinking, especially at the conceptual stage (see Figure 97 in Section 7), and Sanders offers practical advice on creative designing based on his own experience.

(A)

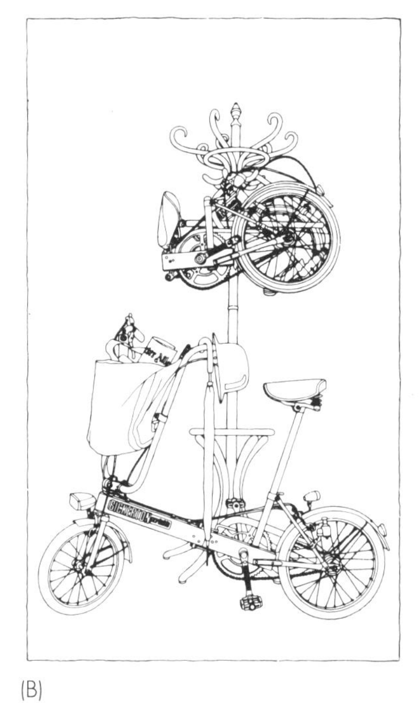

(B)

(C)

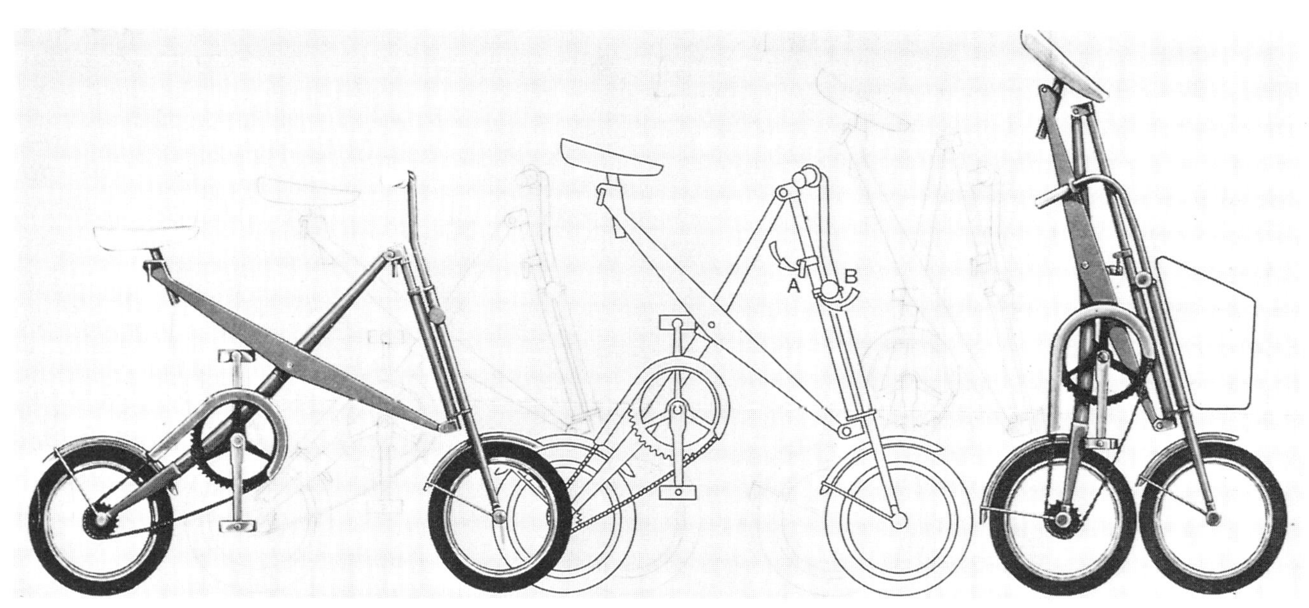

(D)

◀ FIGURE 92
SOME APPROACHES TO PORTABLE BICYCLE DESIGN:
(A) THE STRUTT WORKSONG – A FULL-SIZE BICYCLE WITH A SPECIALLY-DESIGNED FOLD-IN-THE-MIDDLE FRAME TO PROVIDE PERFORMANCE COMPARABLE TO A CONVENTIONAL NON-FOLDING MACHINE

(B) THE BICKERTON – A SMALL-WHEEL BICYCLE WITH A FOLD-IN-THE-MIDDLE FRAME MADE FROM ALUMINIUM ALLOY AND DESIGNED TO BE AS LIGHT AND COMPACT AS POSSIBLE WHEN FOLDED

(C) THE BROMPTON – A SMALL-WHEEL DESIGN WITH A FRAME WHICH HINGES IN THREE PLACES TO PROVIDE A COMPACT FOLDED FORM

(D) BICYCLE DESIGN COMPETITIONS ALWAYS GENERATE LOTS OF IDEAS FOR PORTABLE BICYCLES. THIS SECOND PRIZE WINNER IN A 1973 JAPANESE BICYCLE DESIGN COMPETITION WAS A FOLDING X-FRAME DESIGN

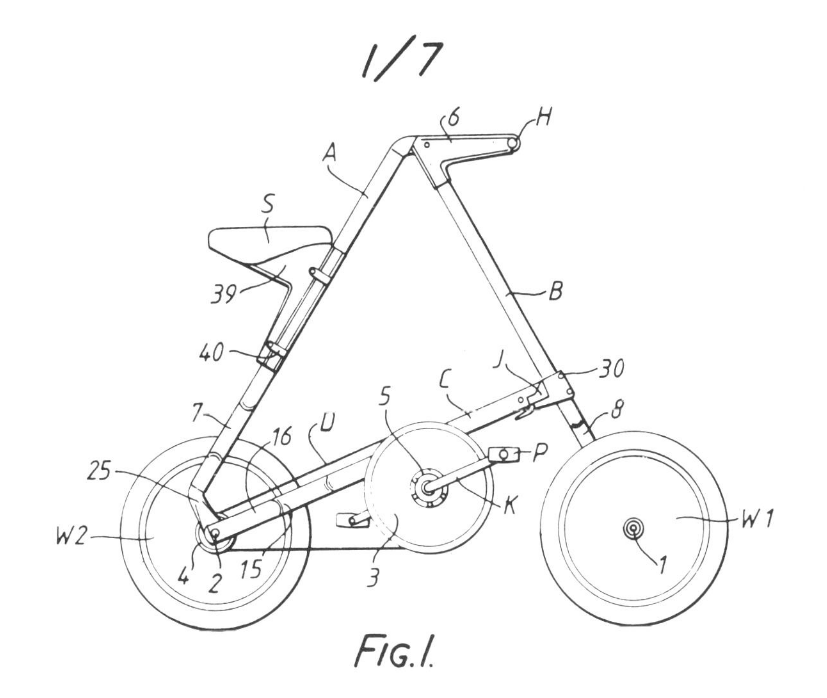

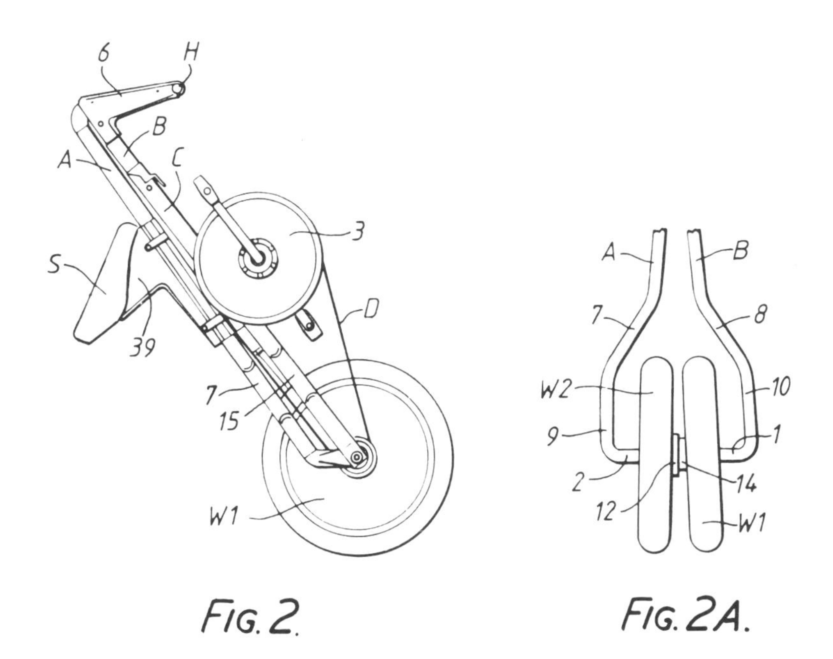

FIGURE 93
THE STRIDA BICYCLE WITH A TRIANGULAR FRAME DESIGNED TO FOLD INTO A LONG THIN FORM THAT COULD BE EASILY WHEELED ALONG THE GROUND LIKE A FOLDED BABY BUGGY. ALUMINIUM AND PLASTICS ARE USED EXTENSIVELY TO SAVE WEIGHT AND THE DRIVE IS VIA AN OIL-FREE TOOTHED BELT

Now look through Section 3 of the *Video 3 Study Guide* and view 'The challenge of the portable bike'. Attempt the self-assessment questions posed in Section 3 of the Study Guide and then try the exercise below. Bear in mind what you have learned when studying Section 7 of this Block and attempting the GDE.

EXERCISE PORTABLE TRANSPORT

A bicycle that folds is the most obvious solution to the problem of designing a portable human-powered transport device. It is the solution adopted by most designers, usually without even considering other options. But a folding bicycle is not the only possible solution. What alternative solution ideas can you think of?

Spend 5–10 minutes noting down your ideas in your Workbook before looking at my attempt below.

You should have noticed that the problem was posed in terms of *function* rather than a particular *solution*, thus allowing a range of alternatives to be considered. Given this broad problem definition, I thought of the following alternatives for a portable human-powered transport device:

folding bicycle	bicycle that takes apart
large-wheel roller skates	skateboard
folding wheelchair	scooter
pogo stick	running shoes

Did you come up with similar ideas or other, perhaps more imaginative, ones? Many of the ideas are likely be impractical but some might be worth developing further, perhaps for special applications for which a folding bicycle would not be suited. In Section 7.3 and on the audio-cassette you will find several techniques to help you generate alternative solution ideas for solving design problems.

7 APPROACHES AND TECHNIQUES FOR INVENTION AND CREATIVE DESIGN

So far we have looked at some theories of creativity, and at how a number of inventors and designers of bicycles and other products came up with their ideas. You've probably wondered how these theoretical ideas and specific examples are going to help you to improve your own creative thinking. Well in this section we turn to some of the approaches and techniques that you can use to undertake creative problem-solving in design. The main purpose of this is to get you thinking about the Guided Design Exercise (GDE) for this Block. So I suggest that you begin by looking again at TMA 03 and the GDE as they are set out in Supplementary Material. You will see that the GDE involves applying a number of approaches and techniques for generating product ideas and developing design concepts. Most of these methods are outlined in this section. You should get a feel for what the various approaches and techniques are and how they can help individuals and groups to produce ideas. Specific techniques are taught in more detail on the audio-cassette and in the *Audio 2 Study Guide*.

Before going on, find and look through the TMA and the associated Guided Design Exercise for this Block.

You will have noticed that I have used the terms 'approaches' and 'techniques' to describe the methods outlined in this section. This is to distinguish between two types:

Conventional approaches for generating ideas and design concepts. These are the methods used intuitively by experienced inventors and designers, often without being aware of what they are doing. Often advice on using such approaches will seem either like simple common sense or like an ideal that is difficult to achieve in practice.

Creative problem-solving techniques are attempts to formalise the processes involved in some aspects of creative thinking. Easily the best known of these techniques is brainstorming, but there are literally dozens of others. Creative problem-solving techniques have their critics and limitations. But, as you will see, they are soundly based on theories of creativity and are undoubtedly useful in helping some individuals and groups to analyse problems and generate ideas, in design as well as in other fields.

7.1 CONVENTIONAL APPROACHES TO INVENTION AND CREATIVE DESIGN

THE CREATIVE PROCESS

As you saw in Section 3, many important inventions and new design ideas arise through the creative process of *preparation*, *incubation*, *illumination* and *verification*. In fact this is the most natural and general approach to creative problem solving in many areas, not just invention and design. The principle behind it is that, provided the mind is well enough prepared with all aspects of a problem, a possible solution or new insight will often occur spontaneously, usually at unexpected moments, due to subconscious patterning of information in the brain.

The secret is in the preparation. Proper preparation requires exploring and defining the problem to be solved, gathering as much information about it as possible, lots of hard thinking, and often making several unsuccessful attempts at a solution. On the video Mark Sanders describes this as 'immersing yourself in the problem'. The idea is to saturate the mind with the problem and possible routes to a solution.

Gordon Glegg, a creative engineering designer, explains:

> Rarely do inventions fall from the blue: they have to be conjured from the conscious and subconscious mind. You cannot command your mind to invent something, but you can encourage it. The best way is to saturate your mind with all the elements of the problem. Study everything you can; try to get the feel of the job.
>
> [...]
>
> The secret of inventiveness is to fill the mind and the imagination with the context of the problem and then relax and think of something else for a change ... If you are lucky, the subconscious will hand up to your conscious mind, your imagination, a picture of what the solution might be. It will probably come in a flash, almost certainly when you are not expecting it.
>
> [...]
>
> Concentration and then relaxation is the common pattern behind most creative thinking.
>
> (Glegg, 1969, pp. 18,19)

This is the approach I used when working out the structure and content of this Block. I tried to define the aims of the Block, read a lot of material, made many notes and attempts at the structure and spent much frustrating time thinking and worrying about the content. The idea for how to structure the Block did not come when I was thinking about it, but while I was taking a shower. It came as an image of the whole thing which I had to write down immediately I got dressed, before it faded from my mind.

Of course it does not end there. Once the idea or solution concept has been 'seen', there remains the process of putting it into practice – through making prototypes, testing, modification and so on until a satisfactory working product can be made. In my case this involved writing drafts of the Block, getting comments on them and modifying the material in response to the feedback. This is often the hardest part, as putting something into practice inevitably throws up new problems to be solved. In fact writing each section of the Block required me to go through another preparation, incubation, etc. sequence with more information gathering, reading, rough notes and hard thinking. But often I found that the way forward came to me not when sitting at my word-processor or consciously thinking about it, but when lying in bed half asleep or while shaving or showering. Of course difficult design problems are not always solved in this way. James Dyson describes on the video the sheer hard slog of getting his ideas to work, and claims he hardly ever gets 'brainwaves in the bath' – for him solutions usually come when 'welding or hammering something in his workshop'. Mark Sanders tends to get ideas for moving forward on a design problem through sketching.

William Middendorf, drawing on his twenty-five years experience of teaching invention and product design, has listed the steps involved in this general approach to creative problem solving:

> 1 Define the problem.
> 2 Gather information.
> 3 Repeatedly review elements of the problem.
> 4 Try for a construction which works even if it is not as elegant as you hoped for.
> 5 Try for an unusual and elegant construction even if it does not satisfy all specifications or work perfectly.
> 6 Repeat steps 1 to 5 several times.
> 7 Direct your attention from the problem ...
> 8 Get back to the problem when you feel enthusiastic about working on it. This may be a signal you are ready for some creative insight.
>
> (Middendorf, 1990, p. 170)

It is worth expanding on the first two, deceptively simple sounding, steps in Middendorf's list.

Problem definition; As you saw earlier in the plate rack example in Section 3, defining the right problem to solve is crucial to coming up with a creative solution. There are a number of approaches and techniques for problem definition, some of which were discussed in Block 2 with respect to product evaluation and drawing up a design brief. I shall not go into this subject any further here except to say that one key to good problem definition is to think in terms of *function* rather than solution. What is the main function of the solution you seek? A simple example of this comes from the way one design consultancy firm approaches problems. When given a brief of designing, say, a dustpan and brush they would always 'start from the premise that we want to get dust and dirt off the floor'. Redefining the problem in terms of function allowed the designers to rethink the familiar dustpan and brush, giving it a palette-shaped handle which can be used to scrape as well as sweep.

> Suppose you were asked to design a new type of vacuum cleaner to a certain specification. That is fine as a problem definition so long as you want to end up with a vacuum cleaner. How might you redefine the problem in more fundamental terms to permit alternative solutions to be considered?

> You need to think 'what is the main *function* of the vacuum cleaner?' Answer: to collect dust and dirt. Then you are no longer constrained to think just of vacuum cleaners, but the field opens to include any device, machine or system that can collect dust and dirt.

The functional approach can work at the component level too. For example, you could consider 'what is the function of the electric motor on the vacuum cleaner?' Answer: to provide rotary power for the fan. Then you are free to think of different methods of providing this power, such as a pneumatic or hydraulic motor, foot pedals, and so on. You could do the same for the fan, the dustbag and the other parts. Functional thinking is the basis of several of the creative problem solving techniques outlined in Section 7.2.

Gathering information; Sometimes an experienced designer will know enough about a particular field to be able to solve a problem without any further information other than that provided in the brief. And, as I mentioned earlier, designers often produce solutions on the basis of very incomplete information.

However, except perhaps in straightforward design problems, it will be almost always be necessary to gather information, either to provide the material for the designer's mind to work on or to verify and refine the designer's initial solution idea.

It is impossible to be sure in advance what information will prove relevant to solving a particular problem. You have seen that often the most creative ideas come from putting together information from previously unrelated fields. So in theory anything could be relevant. But of course it isn't possible to find out everything there is to know to solve a single problem. So where do you start?

Often a designer will start by reviewing the 'state of the art' in the field in which he or she is working. This means looking at existing solutions in the same or similar areas, perhaps by looking through manufacturers' catalogues, visiting shops or showrooms, reviewing patents, reading appropriate books, magazines, trade journals and so on.

Each individual will have their preferred sources of information and inspiration. For example, Tim Hunkin, a well-known inventor-designer, says that when working on a problem 'he reads a lot', especially encyclopedias and catalogues, and may scan past solutions through museum visits. Mark Sanders explains on the video that, as part of the process of 'immersing himself in the problem', he looked at as many publications showing different folding bicycle designs as he could find. Essentially what these designers are doing is to expose themselves to many ideas and principles that might provide them with a solution.

Consulting other people for ideas and information is another very natural thing to do. Most likely these will be experts or other designers in the field. But users, sales-people, retailers or suppliers can also provide very relevant information. Finally, observing existing products in use and becoming a user oneself (perhaps applying the User Trips technique presented in Block 2, Section 2) are good ways of immersing oneself in a problem as well as being valuable sources of information and ideas. As James Dyson says on the video about how he designed the Ballbarrow, 'most of the faults with most products are fairly obvious to any user, so I took each one and tried to get rid of it'.

Let us return to the problem of redesigning a vacuum cleaner to a given specification. What questions might you set yourself to inform yourself about the problem and provide ideas for a solution?

Here is a checklist of possible questions:

1 Where can I find written material which could provide ideas and information?

2 Who can I speak to for ideas and information?

3 Where can I go to to see and use things which might give me ideas and information?

4 What technical or other information do manufacturers have?

5 How can I obtain ideas and information from sellers, purchasers and users?

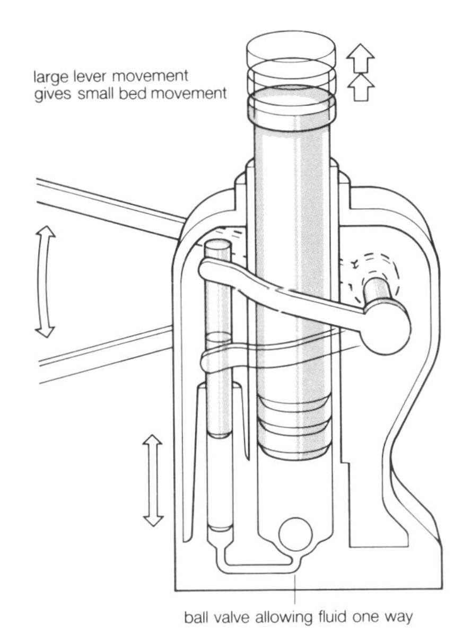

FIGURE 94
ORIGINAL HYDRAULIC JACK USED TO VARY THE HEIGHT OF AN ADJUSTABLE HOSPITAL BED

But don't forget that just collecting information in the hope that a solution will occur 'spontaneously' is not enough. It is necessary also to make *deliberate attempts to create solutions.* Even if these are unsuccessful, they are an essential part of preparing the mind to generate the 'real' solution. So let us look next at some ways of creating solution ideas.

ADAPTATION AND ASSOCIATION

If a problem to be solved is not in a completely novel area, a solution may often be found by finding something similar and adapting it. This common sense approach is intuitively used by many designers.

For example, it was used by one designer who had been given the problem of redesigning a hydraulic lifting device for an adjustable bed to make it more reliable. The device is shown in Figure 94. He looked around libraries and scanned his memory for other devices which lifted things, such as various screw and hydraulic jacks used in industry, and devices which performed similar functions, like passenger lifts and vehicle braking systems. Eventually he realised that the simplest solution would be a system which employed a lever and prop plus locking plates mechanism to move a bar in small controlled steps (Figure 95). He got this idea from a jacking device used in civil engineering which employed the locking plate principle and the use of a locking plate on another part of the bed's tilt mechanism (Dagger and Walker, 1988).

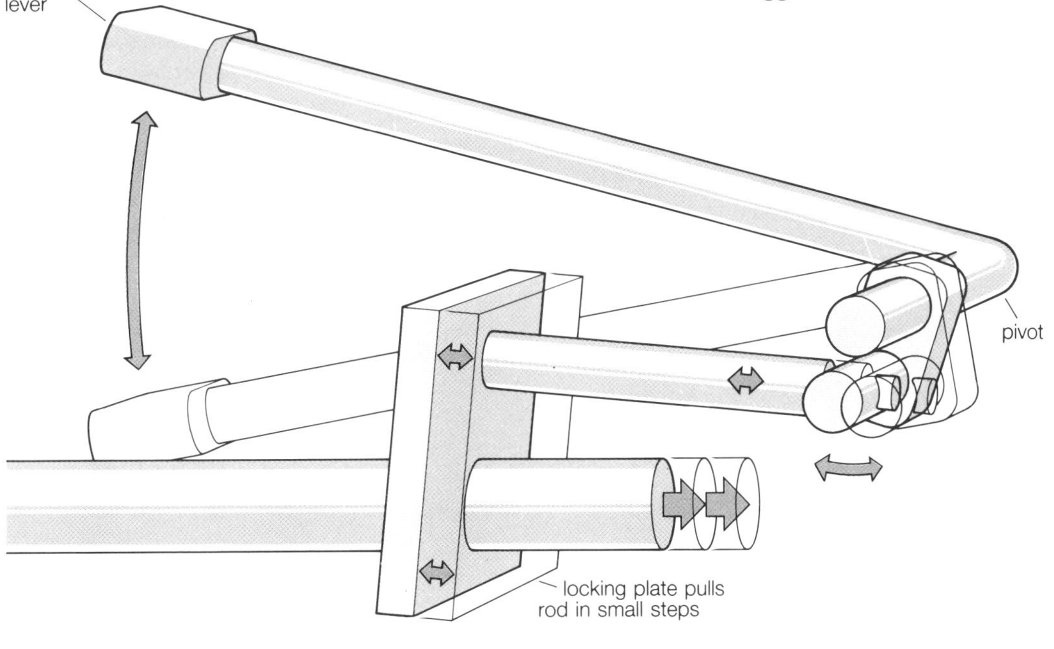

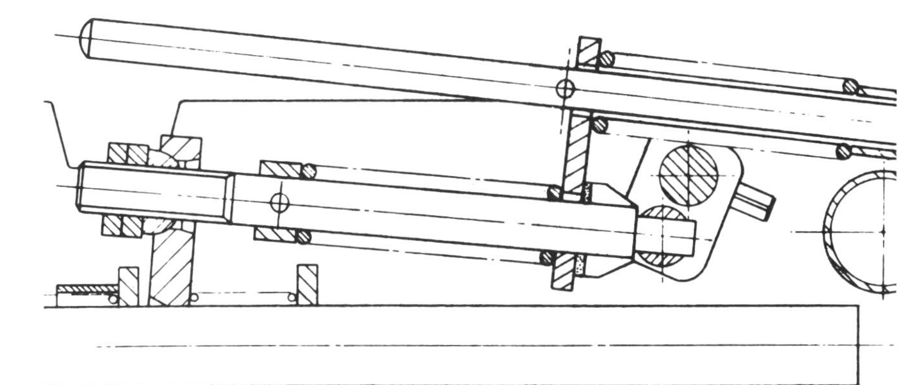

FIGURE 95
NEW MECHANICAL LOCKING PLATE DEVICE WHICH OPERATES THE MECHANISM FOR ADJUSTING THE HOSPITAL BED HEIGHT

Supposing, like the designers in Video 3, Section 3, 'The challenge of the portable bike', you were faced with the problem of designing a portable human-powered vehicle. You have decided to approach the problem in a relatively conventional way by designing some form of folding or collapsible bike. What questions might you pose yourself in order to find some new ideas for the design? What solution principles might these suggest?

Here is another checklist of possible questions:

1. Do I know of a problem that is similar in any respect?
2. What objects or products do I know of which may be folded or collapsed?
3. Do I know of any device which has a similar function?
4. Is there already a device being made and used in a different application which could be adapted for my purposes?

Considering these questions provided me with the following products which function on principles that might be adapted or provide ideas for the portable bike design: baby buggies; umbrellas; folding doors; deckchairs; folding chairs; hand fans; telescopic devices; adjustable ladders; tents.

I think you can see that using a list of questions such as this produces not just ideas for objects that might merely be adapted to produce a new solution, but also helps the designer to think in terms of **functional principles** and stimulates more remote associations between different fields of design and technology.

As you should recall from Section 3, associative thinking of various types is one of the most common sources of inventions and new design ideas. You therefore need to be on the lookout for and be open to ideas and information from almost anywhere that might lead to a solution.

Returning to our vacuum cleaner and portable bike examples. Apart from starting with similar situations and designs, what other sources might lead to an idea for a new product?

Here is my checklist of questions for some of the more likely sources:

1. Can new technologies (e.g. micro-electronics) be applied or provide any ideas?
2. What about new materials (e.g. carbon fibre) or components (e.g. flat 'pancake' shaped electric motors)?
3. Do I know of any scientific discoveries or principles of engineering or technology that may lead to a new solution?

THE PRIMARY GENERATOR

In previous sections you saw that inventors' and designers' solution ideas often arise from a particular dominating constraint, stimulus or image. Industrial, product, graphic and fashion designers in particular use these 'primary generators' as a powerful source of ideas and inspiration. For example an industrial designer may be inspired by an image from a film, a painting, a building, a car or an item of clothing when designing a consumer product. In addition such designers often seek inspiration from trends in society, life-styles, technology, design or fashion. For example, a designer of cookers might think about the following trends:

- in *society* towards working women;
- in *life-styles* towards eating out more often;
- in *technology* towards smaller and more powerful machines;
- in *design* towards automatic or maintenance-free products;
- in *fashion* towards rounded shapes, pastel colours, or whatever.

Any of these trends might provide the main inspiration for a new design of cooker. Does this checklist suggest anything to you? You should recall from Block 2 that some Japanese companies even employ teams of industrial designers aided by sociologists and psychologists to think in precisely these ways.

EXERCISE CHECKLISTS

Before reading on, spend a few minutes noting down, each on a separate page of your Workbook, the four checklists given above in Section 7.1. These checklists should help stimulate ideas when tackling the GDE for this Block. If any of the questions or trends in the checklists prompt ideas now, note these down too. Remember also the advice given in Section 3 about keeping a section of your Workbook for noting 'spontaneous ideas' as and when they occur.

DISCUSSION AND TEAMWORK

Another obvious and common-sense thing to do when attempting to solve a problem or create new ideas is to involve other people. The involvement can range from the occasional discussion to tackling the whole problem in a team. 'Two heads are better than one' goes the old saying; and three heads may sometimes be better than two. Think for example of the highly creative partnerships in popular music and comedy writing.

In invention and design too other people can be used in various ways to help produce and develop ideas. At the beginning of the process they can help to clarify a problem and generate some initial ideas through formal meetings or informal discussion. Many design teams use a loose form of brainstorming 'to toss ideas around' at the early stages of a project (I will describe formal brainstorming later under Techniques). When some initial ideas have been produced, meetings, discussion, and more informal brainstorming is a very good way of generating further ideas and developing the initial ones.

Some individuals use the approach of producing a 'cockshy' effort at a solution simply to provoke the useful flow of ideas that will usually come from others tearing it to pieces! Not being thick-skinned enough to adopt this method, I find that the most useful way to use other people to help me solve a design or other problem is to produce some preliminary ideas and possible solutions on paper, and then to arrange a meeting with one to three other interested people to discuss my ideas. Usually this will produce either a refined or modified version of one of my ideas or a completely new idea, perhaps the result of combining my ideas with those of someone else or comments from another individual sparking off a new train of thought. Really this is only using the combined thinking of several people to stimulate the creative processes of preparation, combination and association. Maybe then I will go away and continue to work on the problem alone, or one of the others will join me to tackle all or part of it.

The point is that by involving other people you open yourself to a wider range of ideas and so you can usually do better than by keeping the problem to yourself. The main danger of this approach is that you might expose your ideas to criticism at a stage when they are only partially formed and vulnerable. One way of guarding against this is to resolve to treat criticisms as opportunities for improving your ideas rather than as a source of discouragement. Nevertheless, it is important to ensure that the people you discuss ideas with are not inclined to be over critical and are prepared to work on the ideas rather than wanting to knock them down.

Think of a time when you successfully solved a problem or created something new. Did you work alone or with one or more others? If you successfully worked with others what kind of people, with what skills or knowledge, were they? Were there some aspects of the task best done alone and others that progressed more successfully with other people?

DRAWING AND MODELLING

Drawing and other forms of modelling are used by designers not just to represent and communicate the ideas and concepts they have envisaged in their heads, but to explore, develop and test out those ideas in a form that is quick and easy to change. Almost all inventors and designers keep books in which they make notes and sketches of their ideas. Figure 96 shows a page from one of Leonardo da Vinci's notebooks and Figure 97 a page from the many sketchbooks used by Mark Sanders when creating the Strida folding bicycle shown on the video.

Of course we are not expecting you to be Leonardo da Vinci or even Mark Sanders. But you can use one of their most useful thinking tools: simple paper and pencil. This is why we encourage you to keep a Work File when tackling your GDE, not just to collect information, but in which to make rough notes, sketches and diagrams when creating design ideas and concepts. Figures 98 and 99 show sketches produced by two students of the previous second-level design course, T263, when thinking of possible improvements to the bicycle. Remember you are using sketches and drawings to aid *your* thinking and develop *your* ideas, so the standard need not be high – the main thing is to represent your ideas visually.

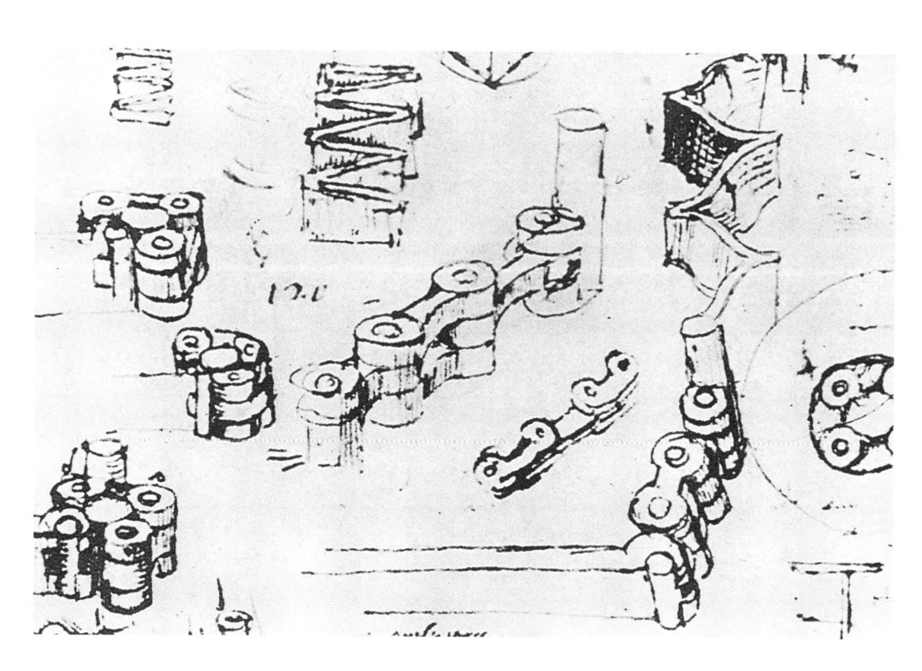

FIGURE 96
A PAGE FROM ONE OF LEONARDO DA VINCI'S NOTEBOOKS SHOWING IDEAS FOR CHAIN LINKS, PROBABLY INTENDED FOR A WHEEL-LOCK GUN

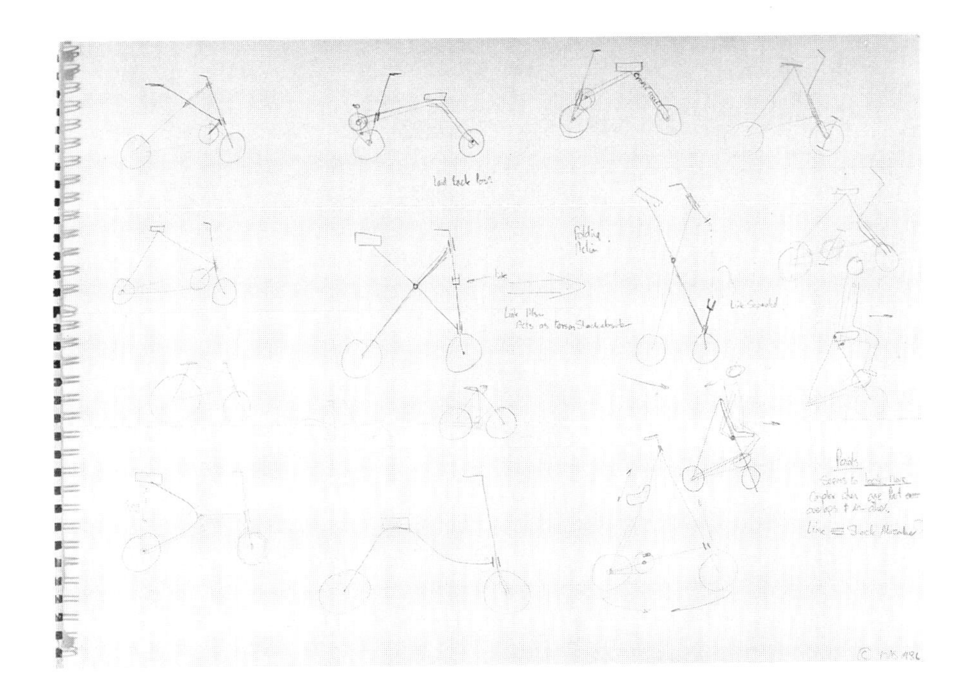

FIGURE 97

A PAGE FROM ONE OF THE EIGHT LARGE SKETCHBOOKS USED BY MARK SANDERS WHEN DEVELOPING THE STRIDA FOLDING BIKE. IT SHOWS SOME DESIGN CONCEPTS FOR ACHIEVING THE LONG THIN FOLDED FORM SANDERS WANTED

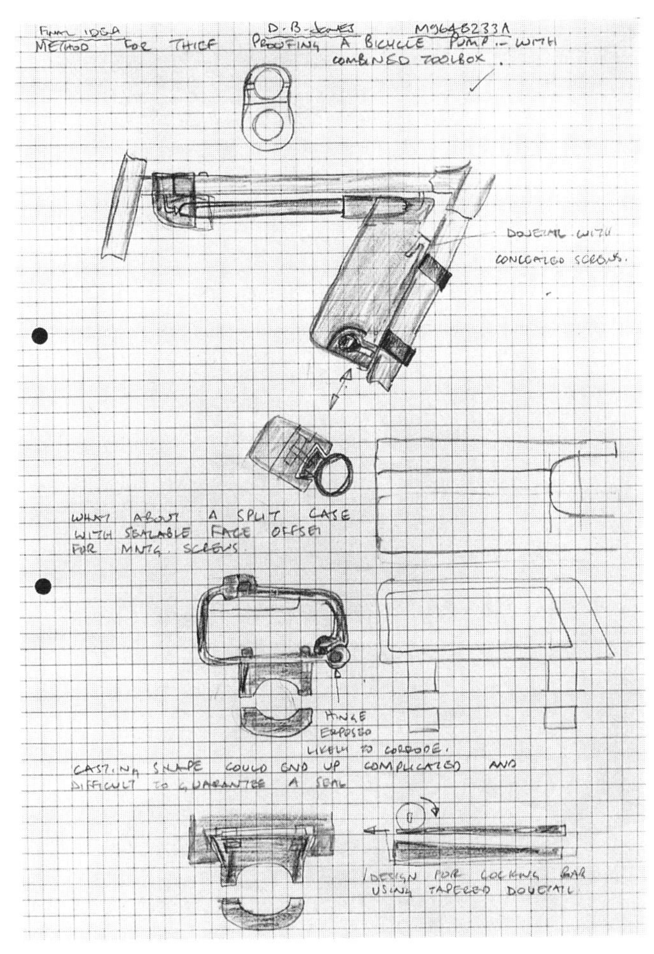

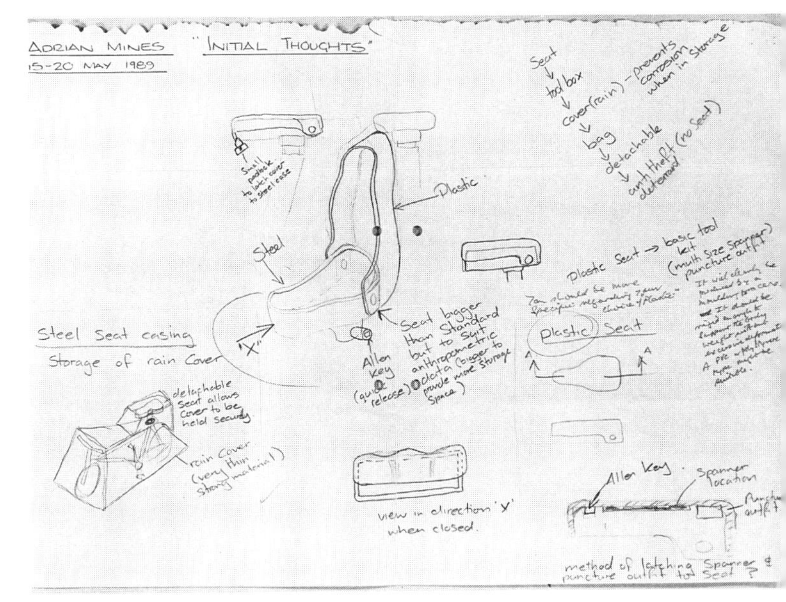

FIGURE 99

PAGE FROM T263 STUDENT'S NOTEBOOK EXPLORING IDEAS FOR A SADDLE FOR STORING A RAIN COVER AND OTHER ITEMS

FIGURE 98

PAGE FROM T263 STUDENT'S NOTEBOOK EXPLORING IDEAS FOR A COMBINED TOOLBOX–BICYCLE PUMP–SECURITY DEVICE

This is a good time to refresh your memory of Section 2 of Video 1 'An introduction to drawing', or to view it for the first time if you have not done so already. Make sure you allow yourself time to practise the drawing exercises before starting serious work on the GDE for this Block.

7.2 CREATIVE PROBLEM SOLVING TECHNIQUES

The conventional approaches to invention and creative design are very effective, as should be evident from the examples I have given so far. But what do you do if you are stuck for ideas? Or if you haven't time to allow for adequate preparation and incubation and need some ideas quickly? Or if you are attempting to break away from conventional thinking and existing designs? Perhaps you are someone who tends to be convergent and analytical and you find it very difficult to think of divergent new ideas. Well, one answer is to try a creative problem-solving (CPS) technique.

Most of these techniques have been developed in the past forty years, although some were invented in the 1930s. So far they have been more popular among managers and marketers than designers (indeed there is an Open Business School course B882: *Creative Management* which teaches a wide variety of CPS techniques). Nevertheless, several of the techniques were mainly intended for use in invention, design and product development.

A very wide range of CPS techniques has been devised for every stage of problem solving from problem definition to solution implementation, but the largest number are primarily intended for idea generation. One book (Van Gundy, 1988) lists sixty idea generation techniques alone. In this section I shall outline a small selection of the simpler and more useful of the *idea generation* techniques. They all provide guidelines to help an individual or a group formalise one or more of the intuitive or conventional approaches involved in creative thinking. But although they are often presented as rules to follow, they are not intended to be applied rigidly like a recipe. The intention is that the techniques once learned should be used flexibly and adapted to the user's own thinking style. They are best used as 'life-jackets' when one is stuck for ideas rather than as 'straight-jackets' to be forced into (Cross, 1989). Often the main benefit of using CPS techniques, especially in a group, is to provide a situation in which participants feel free to be imaginative and in which individuals learn new attitudes which help them to be more creative. My own experience is that idea generation techniques are most useful when, as a result of using conventional approaches, one has defined a specific question to which alternative answers are wanted. For instance, if you were working on the design of a powered cycle, the techniques are a useful way of quickly generating alternative ideas for power sources.

HOW DO THE IDEA GENERATION TECHNIQUES WORK?

In conventional idea generation the inventor's or designer's mind scans his or her memory, printed information and so on for ideas which alone or in combination suggest a way of solving a problem. Many possibilities will be excluded or ignored and one or more ideas will be selected, improved by further thought, and recorded for more detailed development through notes, sketches, and so on.

The idea generation techniques rely on a similar process except that they force the thinker deliberately to widen the search for ideas, to consider odd or zany ideas that otherwise might have been excluded and to make connections that might have been missed even by a highly creative thinker. This is all done before any attempt is made to judge the quality or viability of the ideas. The techniques require the individual or group to 'suspend judgement' so as not to kill off half-formed new ideas by premature evaluation.

Thus, while the conventional creative process tends to take place largely in the head and to oscillate rapidly between divergent and convergent thought, the idea generation techniques are a way of stimulating divergent thinking and getting lots of possibilities down on paper *before* any attempt is made to converge on a solution.

Although there are many idea generation techniques, all work on a limited number of principles:

Intuitive techniques work by stimulating individuals or groups to come up with ideas spontaneously.

Systematic techniques work by breaking the problem into parts, identifying possible sub-solutions and then generating overall solutions by permutation and combination.

These two basic types themselves operate in two ways:

Free association in which new ideas are stimulated in the mind arising from the experience, knowledge and environment of the individual and by ideas produced by other people.

Forced relationships where new ideas are produced in the mind by deliberately forcing together two or more related or unrelated ideas.

Combining these categories gives us four types of idea generation technique, as shown in Figure 100. Within the matrix are the eight idea generation techniques you will find outlined in the the rest of this section.

	INTUITIVE	SYSTEMATIC
FREE ASSOCIATION	Brainstorming Brainwriting Checklists	Attribute listing
FORCED RELATIONSHIPS	Analogies Random stimuli	Morphological analysis Morphological matrix

FIGURE 100
CLASSIFICATION OF IDEA GENERATION TECHNIQUES. (BASED ON GESCHSKA *ET AL.* , 1982; VAN GUNDY, 1988)

7.3 AUDIO: CREATIVITY TECHNIQUES

On Side 1, Part 3 of the audio-cassette for this Block, there is a short discussion of creativity techniques and how they differ from conventional approaches to creativity. You may wish to listen to this part of the audio-cassette now, or wait until you have completed your study of this section.

7.4 EIGHT IDEA GENERATION TECHNIQUES

You are *not* expected to learn the idea generation techniques from the outlines below: just quickly read through to get a feel for what each one can do. Details of how to apply them for the Guided Design Exercise are given on the audio-cassette and in the *Audio 2 Study Guide*.

INTUITIVE TECHNIQUES

Brainstorming

Brainstorming is one of the earliest and certainly the best known of the CPS techniques. It is simply a method of getting a small group of people (ideally five to seven) to generate a lot of divergent ideas in response to a clearly defined problem or question. The difference from informal 'brainstorming' (open-ended discussion) is that there are definite rules, the most important of which is that 'no criticism is allowed' even of the wildest or apparently stupid idea. Evaluation of the ideas comes later. Another important rule is that participants should use the ideas of others to spark off their own ideas and build upon and combine ideas to produce new ones. The outcome of a brainstorming session is therefore a list of perhaps 50 or more ideas for tackling a given problem, or answers to a particular question, which can then be classified and evaluated.

This classical form of brainstorming has proved itself capable of generating useful ideas in areas such as product planning and marketing. For conceptual design a more structured form generally produces more and better ideas. Here brainstorming is used to generate conceptual design ideas, but with the constraints of the brief or specification always in mind – the so-called 'criteria-cued' mode. Alternatively, brainstorming is used to generate ideas for similar products, technologies or situations (rather like the exercise you did in Section 3.3) which may then be used to provide stimuli for new design concepts (Pugh, 1991).

Brainwriting

Brainwriting is similar to brainstorming, except that ideas are written down instead of spoken aloud. It is therefore suited to individual as well as group idea generation. There are several variants of group brainwriting, such as 'Method 635' in which six people each write down three ideas in response to a given problem in five minute intervals. Each person's ideas are then passed to another participant and the process is repeated using the first set of ideas to spark off more. Often the best results are obtained by starting with individual brainwriting and then moving to group brainstorming.

Checklists

A checklist is simply a list of points to be considered or directions to explore when tackling a problem, rather like a shopping list. I have already presented several checklists of questions to be considered in Section 7.1. Many of the items on these checklists are specific to particular types of problem. General CPS checklists have also been developed which attempt to generate new solutions by providing trigger words or other stimuli to thinking about a problem or existing object from new or unusual angles. For example what if the perceived problem were 'reversed' or two existing solutions were 'combined'?

Analogies

As you have seen, analogies are one of the important 'natural' stimuli to creative thinking. CPS techniques have been developed which enable an individual to use analogies more deliberately to obtain new perspectives and insights.

The simplest form involves thinking of something which has some direct or indirect similarity to the problem under consideration (perhaps by brainstorming as suggested above) and then to use this similarity to generate new ideas. More structured techniques focus on the essence of the problem to be tackled (say 'improvement' of a product such as a torch) and use remote or strange analogies (e.g. surgery as being something involving 'improvement') to stimulate new ideas (e.g. the idea of including a emergency tool-kit inside the torch generated by thinking about the instruments needed for surgery). The principle is that odd or even bizarre analogies help to stimulate the mind to think in new ways or jolt it from its normal perspectives (Van Gundy, 1988).

Random stimuli

Random (sometimes called non-logical) stimuli work rather like analogies to stimulate the mind to think in new ways. The stimuli may be produced in a variety of different ways. A simple one is to pick a random word from a book, or picture from a magazine or catalogue, and to think about the word or picture in conjunction with the problem to be tackled to see what ideas it suggests.

Suppose the problem is to devise a way of fastening heavy cupboards to a wall so they are properly aligned horizontally and vertically. The word 'bird' is chosen at random. Thinking about the problem in relation to the characteristics of a bird: flying, feathers, beak, claws, etc. suggests several ideas. For example, birds flying suggested to one individual the idea of an air bag placed underneath the cupboard which by pumping air in or out brings the cupboard into the desired position where it can be easily screwed to the wall (Geschka *et al.*, 1982). Similar devices are already used for positioning very heavy objects.

In theory the more remote the stimulus from the problem in hand the more novel the ideas that might potentially be stimulated. In practice it is often objects or images that are already connected consciously or subconsciously to the problem that work best. Hence other techniques involve switching on the radio or television and using the first word or image that you perceive as the stimulus, or going for a walk and using something in the surroundings that catches your attention as the source of ideas.

Random stimuli, and some of the other intuitive techniques outlined above, may seem silly or far-fetched; but they can and do work, although of course the ideas they generate are not guaranteed to provide the solution to a problem.

SYSTEMATIC TECHNIQUES

The systematic techniques depend on breaking a problem into parts in order to generate solution ideas, and therefore tend to appeal to engineers and others used to a more analytical approach.

Attribute listing

This is one of the earliest of the formal idea generation methods, and is based on the premise that new ideas, inventions and designs arise from previous solutions whose components and their attributes have been modified or changed.

To give a simple, almost trivial, example, suppose you wanted to create a new type of pen. The *components* of an existing pen and their *attributes* of shape, dimensions, material and function, include:

Body Shape: round, 12 cm long by 1 cm diameter; Material: plastic; Function: to support nib, contain ink and provide control.
Nib Shape: round, 1 mm diameter; Material: synthetic felt; Function: to transfer ink from cartridge to paper.
And so on for the **Cap, Ink Cartridge,** etc.

By breaking down the problem in this way it is easier to think of improved designs or new inventions by changing particular components and their attributes. The body shape and size, for example, might be altered to produce one of the short, fat pens now appearing on the market.

Morphological analysis

Morphological analysis is perhaps the best known and most widely used of the systematic techniques. It involves analysing a class of products into its main features or functions (called 'parameters') and then considering different combinations of possible ways of meeting the parameters to generate a very large number of new solution concepts.

My suggested approach in answering the alarm clock exercise in Section 1.3 was a simple example of morphological analysis. In that case the parameters were the 'power source', the 'time indicator' and the 'alarm system'. Different ways of meeting the parameters were identified and combined to generate concepts for different types of clock.

Now consider a more complex example; the different concepts for wave-power devices also given in Section 1.3. Morphological analysis can be used to identify other possibilities. The essential features or functions of any wave-power device is that there has to be a *working surface* on which the waves act, a *balancing force* to hold the device in place, and a way of concentrating the the *wave energy* and transferring it to a turbine or other system for converting it into, say, electricity. By thinking of alternative sub-solutions, or ways of meeting each of these parameters, the morphological chart shown in Figure 101 is produced.

Parameter (Function)	Means of performing function		
Working surface	Flexible sheet	**Rigid solid**	Air surface
Balancing force	Fix to sea bed	Inertia	**Moving structure**
Energy transfer	Air flow	Sea water	**Hydraulics**

FIGURE 101
MORPHOLOGICAL ANALYSIS OF WAVEPOWER DEVICES. (ADAPTED FROM FRENCH, 1988)

The different combinations of sub-solutions in the chart generates 27 ($3 \times 3 \times 3$) possible types of wave-power device, including all those I considered earlier. Another concept suggested by the highlighted combination on the chart is the hinged raft type shown in Figure 102. In fact such a design has been proposed by Christopher Cockerell, inventor of the hovercraft.

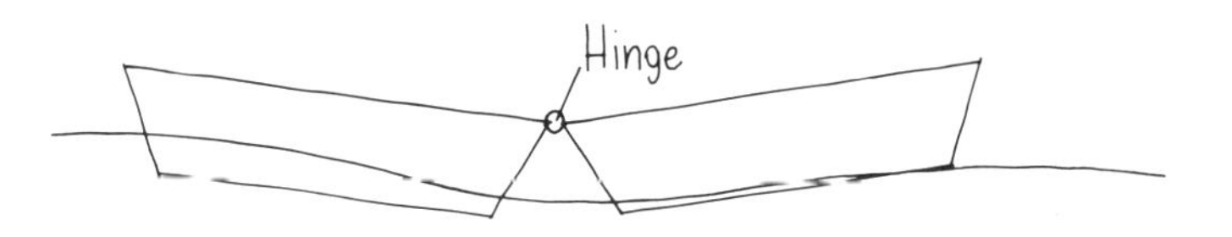

FIGURE 102
HINGED RAFT TYPE OF WAVE-POWER DEVICE. THE WORKING SURFACE COMPRISES THE TWO SOLID HALVES OF THE RAFT WHICH FOLLOW THE WAVE CONTOURS. THE BALANCING FORCE IS PROVIDED BY A STRUCTURE OF LINKED RAFTS AND THE WAVE ENERGY IS TRANSFERRED TO TURBINES BY HYDRAULIC PUMPS OPERATED BY THE RELATIVE MOTION ABOUT THE HINGE

Don't worry if you did not quite follow this rather complicated example. I included it to show that some techniques may be used to produce solution ideas for advanced engineering problems as well as simple design problems.

Morphological matrix

This is a variation of morphological analysis in which just two parameters or dimensions of a problem are chosen and the alternatives for each are listed along the axes of a matrix. This allows systematic forced relationships to be made between the various alternatives. This technique can be used to explore technical options like wave-power devices, but is especially suited to identifying possible new product ideas or market opportunities.

For example, suppose a manufacturer is looking for ideas for new products in a particular field such as refrigeration. A morphological matrix may be drawn up which allows systematic relationships to be made between alternatives for *what* might be cooled (e.g. food, drinks, people, plants, etc.) and *where* the cooling might be needed (e.g. kitchen, bathroom, dining room, car, garden, office, etc.). This will generate some ideas for products that already exist (e.g. cooling food in kitchens) plus others which might be worth exploring further (e.g. cooling drinks in gardens).

7.5 EVALUATION AND SELECTION OF IDEAS

The principle behind the idea generation techniques is to produce as many ideas as possible without any attempt to evaluate them. Often conventional approaches to creative design will also produce several alternative ideas and concepts. To go any further in the design process it is of course necessary to select one or more promising ideas for further exploration and development. This can be done by the intuition or 'gut feel' of an individual or a group, or more systematically using one of the various techniques available for idea evaluation and selection. This is a whole new topic and so I do not intend to detail the techniques here, except to say that most depend on *classifying* ideas to find patterns and/or using checklists of *criteria* (usually those given in the design brief or specification, as discussed in Block 2), to choose between alternatives. The car horn example in Section 1.4 used the latter method. Another useful evaluation technique is known as 'Itemised Response'. So as not to eliminate potentially fruitful ideas too soon, this technique involves looking first at the *positive* aspects of a new idea before considering its weaknesses and the ways these might be overcome (see Rickards, 1982).

Of course the selection process does not end there. Often it will be necessary to go back to the list of ideas, or even generate new ones, if on further examination a selected idea turns out to be technically infeasible, not economically viable or for any reason impractical. And earlier ideas may offer solutions to the new problems that are inevitably thrown up as an idea is developed through the stages of the design process. So no idea should be completely forgotten, even if it is eliminated at an early stage. This is one of the reasons why you are encouraged to keep a Work File for your GDE to record your thinking and the ideas you produce as you go along.

7.6 AUDIO: IDEA GENERATION TECHNIQUES

On side 2 of the audio-cassette for this Block you are guided through a number of exercises provided in the *Audio 2 Study Guide* to help you learn how to use the idea generation techniques outlined in Section 7.4. Side 2, Part 1 of the audio-cassette is concerned with the *intuitive* techniques, and Side 2, Part 2 with the *systematic* techniques. The Study Guide also includes written instructions for carrying out most of the techniques. These are to remind you when applying the techniques in the GDE.

You may wish to use the audio-cassette and the Audio 2 Study Guide to practise the idea generation techniques now, in which case you will need to set aside 1.5 to 2 hours for each of the two parts. Alternatively, you may prefer to learn the techniques as required for the Guided Design Exercise in TMA T264 03.

SAQ 16

Give three ways in which the mental processes involved in using idea generation techniques differ from those involved in conventional creative thinking.

SAQ 17

Classify the following idea generation techniques by their principle of operation:

- Brainstorming
- Random stimuli
- Morphological analysis

8 GOOD IDEAS ARE NOT ENOUGH

So far you have been learning mainly about how inventions and new designs are created from the perspective of the individual. But by now I am sure you are aware that being creative and producing new designs and inventions are not enough to ensure that these become successful new products or innovations in widespread use.

Can you think of examples from the Block which illustrate this important point?

Examples you might have mentioned include: the hovercraft, the pneumatic tyre, xerography and the Moulton bicycle.

What these and other examples show is that there is much work to do, and usually many obstacles to be overcome, before an invention or new design concept becomes an innovation or a new product on the market. And yet further hurdles usually lie in wait which may prevent the innovation or product becoming commercially successful or spreading into general use.

In this section I want to look at these barriers to innovation and to successful diffusion in more detail. This is not to discourage you from coming up with ideas, but to make you aware of what may be involved in translating those ideas into commmercial products.

On Video 3, Section 2, 'Creativity and innovation', James Dyson talks about some of the problems he experienced in getting his Cyclone vacuum cleaner into production and on to the market. You should have viewed this by now; if not, you should do so during your study of this section.

8.1 BARRIERS TO INNOVATION

The barriers to innovation are those which may prevent an invention or new design reaching the market. Here I shall briefly look at technical, financial and organisational obstacles to innovation.

TECHNICAL BARRIERS

The most obvious technical obstacle to innovation is if an idea is scientifically impossible. Any ideas, however imaginative or creative, which violate natural laws are clearly not going to get very far. This rather basic difficulty does not prevent the Patent Office from regularly receiving designs for perpetual motion machines and other impossible inventions.

Another class of idea which is unlikely to make much practical progress is one which is technically infeasible, because of the existing state of the art in technology, materials or knowledge. Earlier in the Block I gave some examples of inventions that were in advance of their time, such as Thompson's pneumatic tyre.

A third technical barrier is when an inventive idea or concept fails to make it to the market because of problems in the development or detailed execution of the design, perhaps to make it sufficiently reliable or economical to manufacture. A related barrier is where the knowledge and technical capabilities of the individual or organisation which is attempting to develop the innovation is not up to the task.

These latter two barriers are clearly not absolute obstacles to innovation as the first two are, but they probably account for many more failures. Examples from the Block are the many wierd and impractical designs produced by nineteenth century cycle manufacturers and designers, partly due to the fact that most were blacksmiths or practical mechanics rather than trained engineers.

FINANCIAL BARRIERS

Generating ideas, inventions and design concepts is usually the least expensive part of the innovation process. Developing those ideas into a practical device or system is much more costly in money, equipment and facilities. Obtaining finance for development of an idea is therefore a major constraint upon innovation, particularly for the independent designer or inventor.

You may recall the example of James Watt who lacked the skills needed to raise the finance to develop his steam engine. Fortunately he went into partnership with Matthew Boulton who possessed the necessary entrepreneurial qualities. More recent examples from the Block of the difficulties faced by inventors in obtaining financial support to develop their ideas include Cockerell with the hovercraft and Carlson with xerography.

ORGANISATIONAL BARRIERS

Innovation involves disturbance and change and therefore tends to be resisted by organisations or the individuals within them. This accounts for the unwillingness of many established businesses to accept new technical ideas, and especially those from outside the organisation itself. You probably recall Alex Moulton's unhappy encounters with the **NIH ('not invented here') syndrome** when he tried to get Raleigh to make his small wheel bicycle.

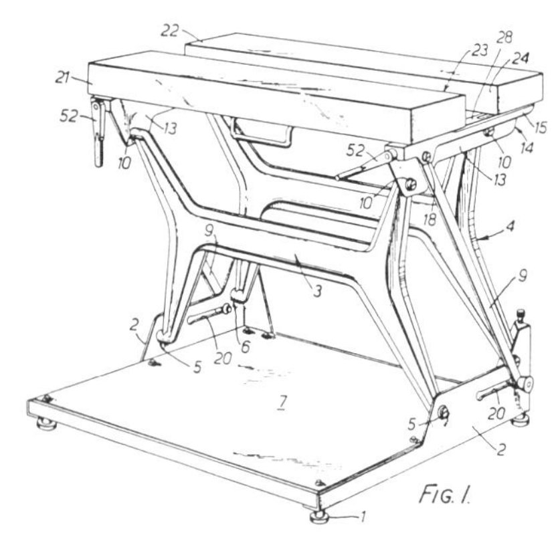

FIGURE 103
PATENT DRAWING OF THE MARK 1 WORKMATE® INVENTED AND DESIGNED BY RON HICKMAN. HICKMAN HAD TO MAKE AND SELL THIS VERSION HIMSELF BEFORE EXISTING MANUFACTURERS WERE CONVINCED THAT HIS INNOVATION WAS COMMERCIALLY VIABLE

A startling example of established organisations failing to see the potential of an innovative design is provided by the Workmate® home workbench (Figure 103), invented by Ron Hickman. None of the existing manufacturers of tools and do-it-yourself equipment that Hickman approached were willing to take on the Workmate, and, like Moulton, Hickman was forced to make and market his innovation himself. Only after the Workmate proved to have a substantial market did manufacturers become interested, and Black and Decker agreed to license the innovation from Hickman – it was to become one of their most successful products, selling over 10 million units in the first ten years (Hickman and Roos, 1982).

Even if an innovative idea comes from within the organisation there are often individuals or departments who will attempt to frustrate its development, especially if it threatens their established skills, traditions or ways of working. An excellent example of this was British Rail's revolutionary, tilting Advanced Passenger Train, which failed to enter full service mainly because internal divisions between BR's engineering departments caused lengthy delays in development which were made worse by internal reorganisation (Potter, 1987). Another factor which delayed development of the APT was the way the project was managed. The APT and many other examples show that design and product development projects, especially innovative ones, usually only succeed in an appropriate organisational structure and with good management. (If you are interested in these managerial aspects they are covered in much more detail in the Open University's third-level course on design and innovation.)

8.2 BARRIERS TO DIFFUSION

So far I have considered the obstacles which influence whether an invention or new design concept becomes an innovation available on the market for the first time. But once an innovation has been introduced, there are many factors which influence the rate at which it sells or spreads into use. These are the barriers to *diffusion*, which depend on several factors: the nature of the innovation itself, of the market for the innovation, and of the individual or organisation promoting the innovation. I am going briefly to examine why some innovations are more successful than others with examples taken from innovations in bicycle design and in information technology.

CHARACTERISTICS OF THE INNOVATION

In his book *Diffusion of Innovations* Everett Rogers (1982) gives five characteristics of an innovation which influence whether and how rapidly it is likely to diffuse. Let's consider these in turn.

Relative advantage

Unless an innovation is perceived to offer advantages over existing or comparable products it is unlikely to succeed. The more highly priced the innovation is relative to comparable goods or services the greater the advantages have to be.

This vital, and obvious, fact is often ignored by inventors, designers and innovators. They are so caught up with the advantages of their new product that they fail to see how it will be perceived by potential buyers and users when they compare it with what else is available. For example, in the 1980s the domestic video-recorder was perceived by millions of consumers to offer them something quite unique at an affordable price. In contrast, satellite television was seen by many British consumers – at least in the early 1990s – as an expensive and unnecessary addition to existing television services.

In general the more mature the product type the more difficult it is to design something that is either uniquely appealing, or which offers a significant advantage over existing goods and services (except perhaps in offering lower prices). That is why it is very difficult to innovate successfully in highly developed products such as bicycles. Lack of sufficient perceived relative advantage, together with some other factors, was the reason for the ultimate failure of the Dursley Pedersen bicycle, and arguably also of Moulton's first small-wheel design. The peformance and comfort offered by these machines was not perceived to be worth their additional complexity and cost. The important word here is 'perceived' advantage. The advantage may be more a matter of image ('vibe' factor in the terminology of Block 1) than significantly improved performance for the price. This is one reason for the great popularity of the mountain bike.

Compatibility

This is the degree to which an innovation is perceived as being compatible with the existing values, skills and past experiences of potential buyers and users. Many innovations fail because they are inconsistent with such values, skills and experiences.

One of the best examples of this is the repeated failure of attempts to improve the standard typewriter or computer keyboard. The existing 'QWERTY' layout was, in fact, deliberately designed to slow down typing speed because early typewriter keys tended to jam if the user typed too fast. Hence there is plenty of scope to design an improved layout and several innovators have tried over the years. Yet none has succeeded, simply because too many people are trained in the use of the standard layout and too many manufacturers are commited to the existing design for a change to be accepted. Even the Maltron keyboard, a novel British design which retains most aspects of the standard layout, but which is ergonomically sculpted to fit the hands, has so far failed to catch on.

Complexity

Complexity is the extent to which an innovation is perceived as being difficult to understand and use. In general, the greater the compexity of an innovation the less likely it is to succeed on the market.

An example is the Microwriter, the world's first fully portable electronic word-processor which was invented by a film producer, Cy Endfield. The Microwriter, which was launched in 1980, employed a unique five-finger keyboard to generate the full range of letters and symbols (Figure 104). However, by 1986 production of the Microwriter had ceased after only some 13 000 units had been sold world-wide. One of the main reasons was that potential users believed the novel five-finger typing method was more difficult to learn than it actually was. Meanwhile it began to lose its unique advantage as portable word-processors with conventional keyboards were developed. Nevertheless, the Microwriting concept was revived in 1989 with the launch of an electronic personal organiser called 'Agenda' which has a dual conventional/five-key keyboard (Figure 105).

Trialability

This is the degree to which an innovation may be tried out on a limited basis. If potential buyers or users have no opportunity to test out the new product or innovation without fully committing themselves to its purchase they are less likely to adopt it.

The importance of trialability in diffusing an innovation was recognised even in the last century. The Singer Company for example offered half-price sewing machines to pastors and church ministers to allow this innovation to be seen and tried out by members of their congregations (Brandon, 1977).

One of best modern examples of trialability is the way the first Xerox photocopiers were introduced. In the late 1950s few businesses were prepared to purchase plain paper photocopiers outright. So the Haloid-Xerox company introduced a scheme in which businesses could lease a Xerox for a limited period for $95. The leasing price included a certain number of 'free' copies, any more were charged per copy. This approach ensured that the Xerox photocopier diffused very rapidly into use and became one of the most commercially successful innovations of all time (Mort, 1989).

Observability

This is the extent to which the results of adopting an innovation are visible to others. In general the more visible an innovation is, and the more it offers its purchaser or user the opportunity to display his or her new acquisition, the more likely it is to succeed. Thus, for example, car telephones as an innovation have been very successful because, as well as providing the unique advantage of being able to contact people while on the move, they offer the owner the opportunity to show off their new 'toy' to friends, colleagues and other drivers. A 'non-observable' innovation, such as a fuel-efficient engine for the car, would therefore tend to diffuse less rapidly.

FIGURE 104
THE MICROWRITER, THE WORLD'S FIRST FULLY PORTABLE WORDPROCESSOR WHICH USED FIVE KEYS TO GENERATE THE FULL RANGE OF LETTERS AND SYMBOLS

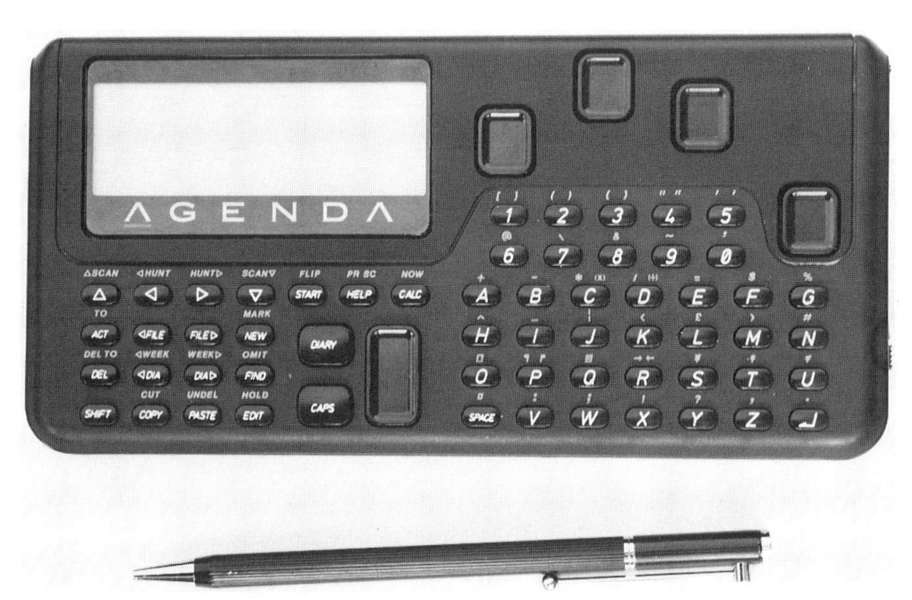

FIGURE 105
'AGENDA' ELECTRONIC PERSONAL ORGANISER WITH A DUAL CONVENTIONAL/FIVE-KEY KEYBOARD

But although high observability is normally associated with rapid rates of innovation diffusion, as you will see shortly, in certain cases potential users may feel they would be embarassed by displaying their purchase and the opposite will apply.

CHARACTERISTICS OF THE MARKET

Apart from the characteristics of the innovation itself, the nature of the market into which it is to be sold has a major influence on diffusion. Some markets contain a high proportion of individuals who welcome novelty. But, typically, the majority of potential adopters of an innovation are conservative and reluctant to adopt anything new. 'Consumer resistance' can be extremely frustrating for the inventor, designer or innovating firm. You may recall the difficulties that J.K. Starley had in getting cyclists to adopt his Rover Safety bicycle and Dunlop's problems in persuading cyclists to buy pneumatic tyres. Both innovations were initially ridiculed by the cycling world, and only became accepted after their superiority had been unquestionably demonstrated in cycle races.

CHARACTERISTICS OF THE INNOVATOR

Thirdly the nature of the innovator, whether an individual or an organisation, can have a major influence on whether an innovation becomes successful.

Kirkpatrick Macmillan, inventor of the first pedal bicycle, was an individual who did not bother to manufacture or sell his machine except in the most limited way. As a result Macmillan's innovation had virtually no commercial impact. Alex Moulton, on the other hand, set up his own business to make and sell his new bicycle, and achieved considerable success before the competition from Raleigh overcame him. James Dyson is successful as an innovator because, at the same time as thinking of ideas for inventions and new designs, he considers their potential for manufacturing and marketing as commercial products.

Likewise innovating firms are only able to succeed in selling a new product if they are able to market it effectively and overcome any resistance from competitors. Sometimes even major companies can fail with an excellent innovation if they are up against competitors with superior market power or marketing ability. Even a company as normally innovative and successful as Sony was forced to abandon its technically superior Betamax video-recorders against overwhelming competition from the rival VHS system.

So far I have outlined some of the many factors that can prevent an invention being developed into an innovation and that innovation diffusing into widespread use. There is one major factor which I have left out which can hamper or promote the development and diffusion of innovations in many fields. What do you think that is? If so, can you give any examples?

The influence of legislation and government policy. These can affect both the creation of innovations and the speed at which they diffuse. For example, as you will see in a moment, the introduction in Britain of new regulations concerning electrically-assisted cycles stimulated several manufacturers to develop such machines. Legislation on limits on environmental pollution from cars are forcing manufacturers to develop cleaner engines and ways of reducing exhaust emissions.

Government policy in France aimed at moving towards an 'information society' provided support to develop the *Télétel* interactive videotex system and to provide large numbers of French users with a free terminal via which they could look up telephone numbers without charge and access a wide variety of other information services. In contrast the British Prestel system, although first on the market, failed to catch on in nearly such large numbers because it relied on individuals and businesses to purchase the necessary equipment and pay for all the services provided.

8.3 TWO EXAMPLES

Let us now test some of these ideas against two examples of innovations in cycle design.

THE SINCLAIR C5

As you know from Block 2, the Sinclair C5 electrically-assisted tricycle was a commercial flop. Following its launch in January 1985 with much promotion and advertising, the C5 sold only a fraction of the 100,000 units per year that its developers, Sinclair Vehicles, had forecast. Production was discontinued in August 1985 after about 5000 vehicles had been sold. Sinclair Vehicles went into receivership in October 1985 having lost of most of the £8.6 million that the C5 project cost, including Sir Clive Sinclair's £7 million personal stake in the venture.

Clearly by risking his own money Sinclair had managed to overcome the barriers to launching the C5 on to the market. But could its commercial failure have been predicted from looking at the barriers to diffusion? Well, let's consider the characteristics of the C5 as an innovation.

Its basic performance specification was set by the 1983 Regulations on electrically-assisted cycles to enable it to be used by anyone over 14 years old without a licence, insurance or helmet. Given this, does it offer a significant *relative advantage* over other modes of short distance transport? Sinclair's view was that the C5 was so different that there was nothing to compare it with – indeed it was marketed as 'a revolution in personal transport'. However, that is not the way that potential users are likely to view it. Once they realise that the C5 is not an electric car, but a vehicle that sometimes needed to be pedalled, most will compare the performance, design and price of the C5 with its closest alternatives: the bicycle and the moped. Is this novel one-person vehicle, with a maximum speed of 15 mph (specified in the Regulations), a limited range, little luggage carrying capacity and initially costing nearly £400, worth buying? Does it seem to offer worthwhile advantages over the alternatives? Other than for those who might be attracted to the C5 for its sheer novelty, the answer is likely to be 'no'. The C5 seems to fail the first test of successful innovation.

Secondly, is the C5 *compatible* with the values, needs and skills of potential users? Sinclair's market research, which asked users for reactions to a prototype, did not seem to indicate problems here. But after the launch it soon became apparent that there were serious concerns about the safety of riding the C5 in traffic, given its low driving position and lack of crash protection (Figure 106). And the actual performance of the C5 (e.g. its range) often did not live up to the advertising. This, together with the poor image the C5 attracted (see below), certainly deterred many potential buyers.

Thirdly, did the C5 seem too complex to learn to use? From my personal experience of riding a C5 I can say that it was quite easy to use, but for those who had not tried the vehicle the prospect of learning may have been a deterrent.

What about *trialability*? The C5 was initially sold by mail order and then through a limited number of outlets, so the opportunities to try out the vehicle before deciding to buy were restricted. However, Sinclair did take this into account to some extent by offering a seven day money back guarantee on mail order sales.

FIGURE 106
THE LOW RIDING POSITION AND LACK OF CRASH PROTECTION WAS ONE OF THE FEATURES OF THE C5 ELECTRICALLY-ASSISTED TRICYCLE WHICH DETERRED POTENTIAL USERS

Finally *observability*. There is no doubt that the C5 was a highly observable innovation! But unusually this acted against it. This is because the C5 quickly attracted a disastrous image, and became the butt of numerous jokes. To ride a C5 seemed to invite public ridicule. In fact my own experience indicates that a C5 rider tends to go virtually unremarked, but in a society which values power and speed in vehicles, it was fairly predictable that an unusual-looking electrically-assisted tricycle, that many people had perceived as an electric car, was unlikely to catch on.

Sinclair of course was not alone in misjudging the demand for electrically-assisted cycles. Many different designs were developed and launched in the wake of the 1983 legislation, although none nearly as innovative or as expensively developed and promoted as the C5. So far none have made any significant impact on the market. Electrically-assisted cycles, constrained in design by over-restrictive regulations plus the low power/weight ratio inherent to lead–acid battery powered vehicles, simply did not provide the customer with a big enough advantage over existing modes of transport to overcome the natural resistance to innovation.

Although it is easy to be wise after the event, I think that you can see that testing the concept of the C5 against some of the barriers to innovation might have led Sinclair to think again. He might have been wiser to develop a proper electric car, as he had originally planned, instead of an electrically-assisted tricycle. In fact, Sinclair decided first to develop an innovative, lightweight electric bicyle, the 'Zike', which was launched in 1992 (Figure 107).

FIGURE 107
SINCLAIR 'ZIKE' ELECTRIC BICYCLE LAUNCHED IN 1992 AT £499. THE ZIKE HAS A FRAME MADE FROM HIGH-TENSILE ALUMINIUM AND PLASTICS COMPOSITE MATERIAL AND IS POWERED BY A 200 W ELECTRIC MOTOR AND NICKEL-CADMIUM BATTERIES, WITH OR WITHOUT PEDAL ASSISTANCE. THE TOTAL WEIGHT AT 11 KILOGRAMS (24 LBS) IS NO MORE THAN THAT OF A LIGHTWEIGHT CONVENTIONAL BICYCLE

THE ITERA PLASTICS BICYCLE

Let's now see what you make of a much less well-known innovation – the Itera plastics bicycle. Can you assess its likely market success or failure from what you have learned?

Many attempts have been made to design and manufacture bicycles made from plastics because of the lightness and resistance to corrosion of plastics materials and the ease with which they can be formed into virtually any shape. Since the Second World War various designs of plastics bicycle have been developed.

To date the most thorough effort to design, develop and market a plastics bicycle has been made by a Swedish firm founded by two former Volvo development engineers. Their design, called the Itera (Figure 108), arose out of some work they were doing on the design of plastic-bodied cars. The Itera is a potentially important innovation, since it is the result of a major design and development effort costing several million pounds – quite an exceptional sum for the cycle industry – and has had the support and involvement of several major international companies.

The frame is a one-piece injection moulding from a glass- and carbon-fibre reinforced plastics material. The forks, wheels and cranks are moulded from the same material. The frame colours – blue-grey, khaki and rose pink – have been chosen for technical reasons: to hide moulding imperfections and to reflect ultraviolet light that might weaken the plastic. Standard steel or aluminium alloy components have been used for gears, chain, seat pin and brakes. The frame, wheels and front forks have all been tested in Volvo's laboratories and shown to be stronger than their equivalents in steel.

FIGURE 108
ITERA BICYCLE WITH FRAME, FORKS AND WHEELS MADE FROM GLASS- AND CARBON-FIBRE REINFORCED PLASTICS. THE ITERA, DESIGNED AND DEVELOPED IN SWEDEN AT A COST OF SEVERAL MILLION POUNDS, WAS LAUNCHED IN BRITAIN IN 1982

Although plastics composites are considerably more expensive than steel, manufacturing costs in plastics can be lower, which has resulted in a total production cost for the Itera comparable to that for a medium-priced conventional bicycle. The high cost of tooling, however, has meant that only one size of frame could be made economically.

The Itera incorporates several novel design features, including, on some models, an integral lighting system with rechargeable batteries and wiring concealed in the frame; maintenance-free bearings; puncture-resistant tyres; a carrying handle; and a build-in lock for the rear wheel. There are also specially designed accessories for the Itera, including plastics panniers and carrying boxes for luggage; a child seat; and a rain cape that retracts into a container behind the saddle when not in use. Unlike conventional bicycles, therefore, the Itera was designed as a total 'package', rather than as an assembly of parts from different suppliers.

The Itera was first launched in late 1981 and introduced in Britain in 1982 at a price of about £140 for the three-speed model, a price then comparable to that for a medium-quality British conventional machine. The main advantages claimed for the Itera is that it is corrosion- and rattle-free, easy to carry, will fit most adults, and requires little maintenance.

Although the Itera was promoted as a lightweight machine, the standard three-speed model is as heavy as a conventional steel roadster, while the ten-speed sports model is rather heavier than many conventional sports bicycles. Those who tested the Itera when it was introduced in Britain agreed that the standard model is hard work to ride, especially uphill. There is less agreement concerning the rigidity of the Itera's plastics components. Several testers mention that various parts of the standard Itera, especially the handlebars in their original plastic version, flex considerably, giving the steering a non-positive quality. But for ordinary riding the test report in *Cyclist* magazine, July 1982, found the frame to be adequately rigid, once the cyclist had become accustomed to the feel of the machine.

So far as appearance and image are concerned, opinions are again divided. Have another look at Figure 108 and see what you think. Many consider the Itera ugly, looking rather like a nineteenth-century boneshaker, and dislike or distrust plastics as materials for a bicycle. Others seem to be positively attracted to the Itera's novel appearance, to the modernity and cleanliness associated with plastics, and to features like the partially enclosed chain.

The interesting question, therefore, is whether the Itera succeeded where other plastics bicycles before it have failed. That is what I want you to assess in the following exercise.

EXERCISE SUCCESSFUL INNOVATION

On the basis of the barriers to diffusion discussed in this section, and the above description of the Itera bicycle, assess whether the Itera is likely to succeed as an innovation.

Spend ten minutes on this exercise before looking at my answer below.

Consider the innovative Itera bicycle in terms of the various barriers to diffusion.

First, the characteristics of the innovation itself. In terms of *relative advantage* the Itera offers several unique features (e.g. corrosion resistance), but no worthwhile performance, weight or cost advantages over conventional bicycles. Indeed there may be some drawbacks such as flexing of the frame. Regarding *compatibility,* it is similar in basic design to conventional bicycles, but some people may be prejudiced against plastics as a structural material. There should be no difficulties with *complexity* as the Itera is no more difficult to ride and easier to maintain than conventional machines. Regarding *trialability*, no arrangements seem to have been made to try out the machine on a limited basis. Finally *observability*: the novelty and distinctive design would attract some buyers, but the rather ungainly appearance and unattractive colours might deter others.

Second, the characteristics of the market. Although the Itera might well appeal to those who are not already cyclists, existing cyclists and cycle dealers tend to be highly resistant to innovation. It would need to become a fashionable product, like the mountain bike, in order to overcome this consumer resistance.

Third, the characteristics of the innovators. The Itera was developed by a team of professional engineers and designers, backed by several large companies and launched using sophisticated marketing methods. This is certainly a major factor in the Itera's favour.

On the basis of the above it seems likely that, despite the highly professional approach of its designers and manufacturers, the Itera offers too few practical advantages over conventional bicycles and several visual design deficiencies to succeed as an innovation. In fact manufacture of the Itera ceased in 1983, less than two years after its introduction.

8.4 A FINAL WORD

This Block has shown that creative thinking is vital to produce the inventions and innovative design ideas that are the source of the many new products available in a modern industrial society. But you have also seen that good ideas are not enough: creativity cannot ensure that an idea is developed into an innovation or make that innovation diffuse into widespread use.

In this Block I have concentrated on the work of *individual* inventors and designers, from Starley and Dunlop in the nineteenth century to Moulton and Dyson in the twentieth. Although highly creative individuals like these continue to produce innovative products, they are only responsible for a small part of inventive and design work going on today. As you saw in Section 1, most designing is evolutionary rather than innovative. And an increasing proportion of invention and design is undertaken by teams working in research institutions and large companies. Even individual inventors and designers usually have to work closely with others if they are to manufacture and market their products. Promoting creativity in teams and within organisations is something I have only touched upon; it will be taken up again in Block 5. Nevertheless, several of the approaches and techniques introduced in the Block to help you to think more creatively as an individual are equally, and sometimes more, effective for creative problem-solving in teams or groups.

ANSWERS TO SELF-ASSESSMENT QUESTIONS

SAQ 1

Invention: something completely new; a description, sketch or model conveying the essential principles of a new product or process, e.g. Cockerell's hovercraft model.

Design: A particular physical embodiment of an inventive principle e.g. the SR.N1 prototype hovercraft; a new form of an existing product or device, e.g. the redesigned AP-188 hovercraft.

Product innovation: a novel product or device at the point of first commercial introduction or use, e.g. the SR.N6 hovercraft when first launched.

Diffusion: the process of adoption into use of an innovation, e.g. the spread of the hovercraft into passenger service on various routes.

SAQ 2

The product design specification (PDS) is the normal starting point for conceptual design. The PDS defines the problem and the constraints on the search for solutions and provides criteria for evaluating alternative concepts.

SAQ 3

Analytical, creative and evaluative thinking are required in design. Creative thinking is required throughout the design process, but is especially important at the conceptual stage because that is where the alternative solutions to the design problem at its most general level are generated.

SAQ 4

Although it is not possible to be sure, the most likely sources are:

- Draisienne: *adaptation* of the child's hobby horse with the addition of steering;
- Macmillan bicycle: *transfer* of craft knowledge of mechanisms from one application to another;
- Velocipede: the *analogy* between the cranked handle turning a grindstone and a device for turning a bicycle wheel.

SAQ 5

A product which has evolved into a configuration that has displaced all previous designs on the market or in use.

The diamond-frame bicycle took several years to become dominant partly due to the lack of technical understanding on the part of cycle manufacturers and designers and partly due to the conservatism of cyclists. This meant that inferior designs continued to be made and sold even after the appearance of the diamond-frame safety bicycle.

SAQ 6

In many types of problem solving the main stages of the creative process have been found to be: *identification* (of a problem and determination to solve it); *preparation* (exploring the problem and attempting solutions); *incubation* (subconscious patterning by the relaxed mind); *illumination* (the sudden 'flash of insight'); *verification* (testing the idea and developing it into a workable solution).

The theory certainly seems to support Edison's saying, as much effort is involved in all the preparatory activities which may result in the production of a creative idea, and even more effort is usually involved in getting the idea to work in practice. This is especially true of invention and product design where the verification stage usually requires the greatest effort.

SAQ 7

Adaptation: modifying available solutions or technologies to a new application, e.g. use of pistol grip for variety of products.

Transfer: idea or invention stimulated by an innovation in another field, e.g. transfer of laser technology from military to domestic applications. (It is often difficult to distinguish between adaptation and transfer.)

Combination: of existing devices or ideas to produce something new, e.g. combined screwdriver and wire-stripper.

Analogy: a similar situation or problem which suggests a solution in a different area, e.g. plant forms used to suggest structures in bridge design.

All these types of thinking involve making mental associations or links between ideas, knowledge or principles in different areas to suggest a solution to a problem.

SAQ 8

The primary generator is an idea or objective arising from the experiences and preferences of the designer which suggests an initial solution to a complex design problem. Designers often have a primary generator as a way of getting started upon, and making manageable, a design problem with a vast number of possible solutions.

SAQ 9

J.K. Starley's main contribution in the evolution of the bicycle was his dissatisfaction with the riding efficiency of existing designs and creating a configuration which placed the handlebars, pedals and seat so as to optimise the rider's effort.

Starley depended on many prior component innovations in producing his Rover Safety, including chain drive, tangent spoked wheels and hollow steel frames, as well as a variety of cycle designs, such as the Humber tricycle, which had a diamond frame.

SAQ 10

General lessons about invention and innovation from the case of the pneumatic tyre include:

A latent demand for an invention, and suitable materials and techniques for making it into an economic and acceptable product, must exist. Thompson failed for lack of these requirements.

There is often resistance to innovation from manufacturers and users. Dunlop managed to overcome this by demonstrating the superiority of the pneumatic tyre in cycle races. Thompson was only able to demonstrate his tyres in a very limited way.

An innovation may become widely accepted only when its initial deficiencies are removed through further invention and innovation. It was the detachable pneumatic tyre that succeeded, not Dunlop's original tyre.

SAQ 11

Rover Safety bicycle: Starley's dissatisfaction with the ergonomics of existing bicycles. Thinking of the analogy between a man climbing a ladder and riding a bicycle enabled Starley to produce his ergonomically improved design.

Dunlop's pneumatic tyre: Dunlop's long interest in vehicle vibration and the desire of his son for a faster cycle. The origins of the idea are not known, but it seems likely that the air-filled rubber tyre stemmed from Dunlop's work as a vet, probably from his experience of making rubber appliances or possibly from seeing animal intestines.

Dursley Pedersen bicycle: Pedersen's dissatisfaction with the conventional saddle led him to invent a novel hammock-like seat. The bicycle design arose from Pedersen's desire to produce a light and elegant frame suited to his new type of seat.

SAQ 12

My list is as follows. You might have produced a different, or longer list, but it should include most of the items below.

The product or device should be *technically* sound (in mechanics, structure, materials, etc.) and of acceptable *cost*.

If the product interacts directly with human users it should also be *ergonomically* correct (*very* important for bicycles),and of acceptable *appearance*.

In addition, to ensure successful innovation, the inventor or designer must have the necessary *personal qualities* to innovate; the *market* should be favourable; the product should be appropriate to *social* and *cultural* conditions.

SAQ 13

Convergent thinking involves analysing a given problem in order to select a logically correct or expected answer from alternatives.

Divergent thinking involves generating alternative solutions to a given problem. It involves fluency (mental ability to produce many responses to a problem from stored information); flexibility (ability to break out of usual or preconceived thought patterns); originality (ability to think of novel or ingenious solutions to problems).

Both types of thinking are required for creative design: divergent thinking to generate original ideas and alternative solutions and convergent thinking to analyse and select between the alternatives.

SAQ 14

The Moulton bicycle made cycling fashionable and stimulated new design thinking in the industry. It was superceded by rival designs with inferior performance and comfort due to the market power of the major manufacturers and the greater cost and complexity of the Moulton. The Moulton was intended as a universal design while the 'shopper' and 'Chopper' were designed to exploit specific, growing market segments.

SAQ 15

Moulton's design process is systematic involving: deep study of a problem leading to ideas for a solution; sketching and analysing the concepts; making and testing prototypes; modifying the design and developing it for manufacture. The mountain bike was designed by trial and error experimentation with machines built from existing components.

Moulton' s systematic approach is suited to innovative design, whereas the trial-and-error approach is more suited to evolutionary design.

SAQ 16

Idea generation techniques differ from conventional approaches to creative thinking in three main ways:

- They force the thinker to deliberately widen the search for ideas for solving a problem by exposure to a variety of external stimuli – other people's ideas; checklists of questions; random words, forced connections, etc.
- Conventional thinking usually oscillates between idea generation and the evaluation of ideas, whereas idea generation techniques require deliberately leaving evaluation until a later stage. Odd or zany ideas which might prove useful are not excluded by premature judgement.
- Idea generation techniques force the thinker to put down on paper ideas which in conventional creative thinking might remain in his or her head.

SAQ 17

- Brainstorming: intuitive; free association
- Random stimuli: intuitive; forced relationships
- Morphological analysis: systematic; forced relationships

CHECKLIST OF OBJECTIVES

Having completed your study of this Block you should now be able to do the following.

SECTION 1

1 Explain what is meant by **creativity** in the fields of invention and innovative product design.

2 Define and, using examples, distinguish between **invention, design, innovation** and **diffusion.**

3 Explain how **conceptual design** fits into the total product design process, and why creativity is especially important at the conceptual stage.

4 Distinguish between **analytical, creative** and **evaluative** thinking in product design.

SECTION 2

5 Outline how the inventors and designers of the first bicycles (Drais, Macmillan and Michaux) are likely to have got their ideas.

6 Explain what is meant by a **dominant design.** Discuss why there was such a variety of cycle designs in the nineteenth century before the diamond-frame bicycle became dominant.

7 (After viewing Video 3, Section 1, 'The evolution of the bicycle') explain some of the technical and other factors influencing the evolution of cycle designs in the twentieth century.

SECTION 3

8 Outline the main stages of the creative problem-solving process and discuss the importance of **illumination** or the 'flash of insight' relative to the other stages.

9 Explain, using historical examples, how mental association between different areas of knowledge is the source of many creative and inventive ideas.

10 Explain how redefining a design problem in functional terms can overcome mental barriers to creating a new solution.

11 Explain what is meant by the **primary generator** which provides a designer with the initial concept for solving a design problem.

12 Outline what is meant by the following types of **associative thinking:** adaptation; transfer; combination; analogy. Using examples explain similarities and differences between these sources of creative ideas for inventions and new designs.

13 (After viewing Video 3, Section 2, 'Creativity and innovation') discuss, using examples, the role of theoretical understanding and practical experimentation in the production of major inventions.

SECTION 4

14 Outline the main steps in cycle design and component innovation required before the Rover Safety became possible.

15 Discuss why the Rover Safety took over twenty years to evolve from the 'boneshaker' and the factors promoting and inhibiting that evolution.

16 Compare and contrast the sources of the ideas underlying the conceptual design of the Rover Safety and Dursley Pedersen bicycles and the invention of Dunlop's pneumatic tyre.

17 Discuss barriers to the diffusion of an innovation using the examples of the Rover Safety and the pneumatic tyre.

18 Give reasons why the Rover Safety and Dunlop's pneumatic tyre became very successful innovations, while the Dursley Pedersen bicycle and Thompson's pneumatic tyre soon disappeared from use.

SECTION 5

19 Explain why **constructive discontent** and a high motivation to solve problems tend to be personality characteristics shared by inventors and creative designers.

20 Explain what is meant by **convergent** and **divergent thinking,** and discuss the role of divergent and convergent thinking in creative design.

21 Understand the importance for an inventor or creative designer of having an 'inventor's eye' and the skill of 'thinking with the hands'.

22 Understand why sketching and drawing and visualisation of objects are important skills in creative design.

23 Discuss the extent to which specialist knowledge and experience are required for invention and innovative design in a particular field.

24 (After attempting the 'Thinking styles exercises' in the *Audio 2 Study Guide*) become more aware of your own abilities for visualising three dimensional objects, divergent/lateral thinking and convergent/logical problem solving.

25 Appreciate what is involved in attempting simple design problems requiring creative solutions.

SECTION 6

26 Explain the shift from **product** to **process innovation** in the evolution of the bicycle during the twentieth century and (after viewing Video 3, Section 1, 'The evolution of the bicycle') discuss the positive and negative effects of cycle sport on innovation in cycle design.

27 Outline the creative process used by Alex Moulton to invent and design his small-wheel bicycles and identify similarities with the process used by the creators of the Rover Safety and Dursley Pedersen bicycles.

28 Explain why the small-wheel bicycle conceived by Moulton was less commercially successful than its more poorly-engineered successors.

29 Compare the design process used by Moulton with that of the individuals who created the mountain bike.

30 Discuss the relative importance of marketing and design in the popularity of the mountain bike. Give other examples where **psychological** as well as **practical design factors** were important in the success of a cycle innovation.

31 (After viewing Video 3, Section 3, 'The challenge of the portable bike') identify the practices and approaches used by the designer of the Strida folding bicycle with a view to applying them to similar creative design problems.

SECTION 7

32 Understand that creative ideas sometimes occur at unexpected moments and apply that understanding to your own creative work.

33 Use checklists to identify (a) sources of relevant information, and (b) similar problems or situations that might provide ideas to solve a design problem.

34 Appreciate the practical value of (a) discussion and teamwork, and (b) drawing and modelling when tackling creative design problems.

35 Understand the principles behind **creative problem-solving techniques** and how they differ from conventional approaches to creative thinking.

36 Understand the purpose of the following **idea generation techniques:** brainstorming, brainwriting, checklists, random stimuli, attribute listing, morphological analysis, morphological matrix and (after using the associated audio-cassette and *Audio 2 Study Guide*) apply the techniques for solving simple problems in product planning and creative design.

SECTION 8

37 Discuss, using examples, the technical, financial and organisational **barriers to innovation.**

38 Discuss, using examples, the **barriers to diffusion** of an innovation, including the five characteristics of the innovation itself identified by Rogers.

39 Apply the barriers to diffusion to assess the likelihood of a particular product innovation becoming commercially successful.

REFERENCES

Adams, J.L. (1987) *Conceptual Blockbusting: a guide to better ideas*, Harmondsworth: Penguin.

Brandon, R. (1977) *Singer and the Sewing Machine – a Capitalist Romance*, Barrie and Jenkins.

Cross, N. (1989) *Engineering Design Methods*, Chichester: John Wiley.

Dagger, B. and Walker, D. (1988) 'Case studies: Fisher-Miller; Nesbit Evans' (P791: *Managing Design*, Unit 1), Milton Keynes: The Open University Press.

Darke, J. (1979) 'The primary generator and the design process', *Design Studies*, Vol.1 No.1, pp. 36–44.

Dixon, J.R. (1966) *Design Engineering: inventiveness, analysis and decision-making*, McGraw Hill.

du Cros, A. (1938) *Wheels of Fortune*, London: Chapman and Hall.

Dunlop, J.B. (1922) *The History of the Pneumatic Tyre*, Alex Thom.

Dyson, J. (1987) Transcript of talk to The Bath Design Conference 1987 'Designer as Entrepeneur'.

Evans, D.E. (1978) *The Ingenious Mr. Pedersen*, Allan Sutton.

French, M. (1985) *Conceptual Design for Engineers*, 2nd edition, London: The Design Council.

French, M. (1988) *Invention and Evolution: Design in Nature and Engineering*, Cambridge: Cambridge University Press.

Geschka, H., von Reibnitz, U. and Storvik, K. (1982) 'Idea generation methods: creative solutions to business and technical problems', *Battelle Technical Inputs to Planning, Review No.5*, Columbus Ohio: Battelle Memorial Institute.

Glegg, G.L. (1969) *The Design of Design*, Cambridge: Cambridge University Press.

Gorman, M.E. and Carlson, W.B. (1990) 'Interpreting invention as a cognitive process: the case of Alexander Graham Bell, Thomas Edison and the telephone', *Science, Technology and Human Values*, vol.15 (2), Spring, pp. 131–64.

Hadamard, J. (1945) *The Psychology of Invention in the Mathematical Field*, Princeton N.J: Princeton University Press.

Hadland, A. (1982) *The Moulton Bicycle*, 2nd edition, privately published, Reading, Berks.

Hickman, R.P. and Roos, M.J. (1982) 'Workmate', *CIPA* (Chartered Institute of Patent Agents) *Journal*, July, pp. 424–57.

Holt, K. (1987) *Innovation: a challenge to the engineer*, Amsterdam: Elsevier.

Hudson, L. (1967) *Contrary Imaginations*, Harmondsworth: Penguin.

Hughes, T.P (1978) 'Inventors: the problems they choose, the ideas they have, and the inventions they make', *in* Kelly, P. and Kranzberg, M. (ed.) *Technological Innovation: a critical review of current knowledge*, San Francisco: San Francisco Press, pp. 166–82.

Jones, J.C. (1970) *Design Methods: seeds of human futures*, London: John Wiley.

Kelly, P. *et al.* (1978) 'The individual inventor-entrepreneur', *in* Roy, R. and Wield, D. (eds) *Product Design and Technological Innovation*, Milton Keynes: The Open University Press, pp. 76–85.

Kelly, C. and Crane, N. (1988) *Richard's Mountain Bike Book*, Yeovil: Oxford Illustrated Press.

Kirton, M.J. (1980) 'Adaptors and innovators: the way people approach problems', *Planned Innovation*, No.3, March/April, pp. 51–4.

Lawson, B. (1980) *How Designers Think*, London: Architectural Press.

Ledsome, C. (ed.) (1987) *Engineering Design Teaching Aids*, London: The Design Council.

Mackinnon, D.W (1970) 'Creativity: a multi-faceted phenomenon', *in* Roslansky, J.D. (ed.) *Creativity: a Discussion at the Nobel Conference*, Amsterdam: North Holland.

Mansell, C. (1973) 'The rallying of Raleigh', *Management Today*, February, 83–92.

Middendorf, W.H. (1969) *Engineering Design*, Boston: Allyn and Bacon.

Middendorf, W.H. (1990) *Design of Devices and Systems*, 2nd edition, New York and Basel: Marcel Dekker.

Mort, J. (1989) *The Anatomy of Xerography*, Jefferson, North Carolina: McFarland.

Moulton, A.E. (1966) 'Design and technological innovation', paper given to the Design Congress 'Profit by Design'.

Moulton, A.E. (1973) 'The Moulton bicycle', *Journal of the Royal Society of Arts*, pp. 217–34.

Moulton, A.E. (1979) 'Innovation', *Journal of the Royal Society of Arts*, December, pp. 31–44.

Ochse, R. (1990) *Before the Gates of Excellence: the determinants of creative genius*, Cambridge: Cambridge University Press.

Papanek, V. (1972) *Design for the Real World*, London: Thames and Hudson.

Potter, S. (1987) *On the Right Lines? The Limits of Technological Innovation*, London: Frances Pinter.

Pugh, S. (1991) *Total Design: integrated methods for successful product engineering*, Wokingham: Addison-Wesley.

Pye, D.W. (1978) *The Nature and Aesthetics of Design*, London: Herbert Press.

Rauck, M. (1983) *Karl Freiherr Drais von Sauerbronn: Erfinder und Unternehmer (1785–1851)*, Wiesbaden: Franz Steiner Verlag.

Rickards, T. (1982) *Problem Solving through Creative Analysis*, Aldershot: Gower Press.

Roberts, D. (1990) 'The first bicycle', *in KM150: A celebration of the invention of the pedal cycle by Kirkpatrick Macmillan*, Special Souvenir Handbook, KM150 International Cycling Festival, Dumfries and Galloway, May.

Rogers, E. M. (1983) *Diffusion of Innovations*, 3rd edition, New York: Free Press.

Rothwell, R. and Gardiner, P. (1985) 'Invention, innovation, re-innovation and the role of the user: a case study of British hovercraft development', *Technovation*, vol 3, pp. 167–86.

Roy, R. with Cross, N. (1983) 'Bicycles: invention and innovation', (T263: *Design: Processes and Products*, Units 5–7), Milton Keynes: The Open University Press.

Sanders, M. A. (1985) 'The design of a new folding bicycle', Unpublished Masters Thesis, Royal College of Art/Imperial College of Science and Technology.

Saul, S.B (1970) *Technological Change: the United States and Britain in the nineteenth century*, University Paperbacks.

Shahin, M. M. A., (1988) 'Application of a systematic design methodology: an engineering case study', *Design Studies*, vol. 9, (4), October, pp. 202–7.

Sharp, A. (1896) *Bicycles and Tricycles*, London: Longmans Green, republished 1977 by MIT Press.

Starley, J.K. (1898) 'The evolution of the bicycle', *Journal of the Society of Arts*, vol. 46, 20 May, pp. 601–16.

Stein, M.I. (1956) 'A transactional approach to creativity', *in* C.W.Taylor (ed.) *The 1955 Utah Research Conference on the Identification of Creative Scientific Talent*, Salt Lake City: University Utah Press.

Thring, M.W. and Laithwaite, E.R. (1977) *How to Invent*, London: Macmillan.

Van Gundy, A.B. (1988) *Techniques of Structured Problem Solving*, 2nd edition, New York : Van Nostrand Reinhold.

Walker, D. with Cross, N. (1983) 'An introduction to design', (T263: *Design: Processes and Products*, Unit 1), Milton Keynes: The Open University Press.

Walker, D., Dagger, B.K.J. and Roy, R. (1991) *Creative Techniques in Product and Engineering Design*, Cambridge: Woodhead Publishing.

Whitfield, P.R. (1975) *Creativity in Industry*, Harmondsworth: Penguin.

ACKNOWLEDGEMENTS

Grateful acknowledgement is made to the following sources for permission to reproduce material in this Block:

Text

Moulton A. (1979) 'Innovation', Reproduced with permission from the RSA *Journal*, vol. CXXVIII, no. 5281, December 1979.

Figures

Figures 3, 4 and 6: Hovercraft Consultants Ltd; Figure 5: © Crown Copyright; Figure 7: Westland Aerospace Ltd; Figure 10: Shahin M. (1988) *Design Studies*, October 1988, Butterworth Heinemann; Figure 12: Pugh S. (1990) *Total Design*, Prentice-Hall College, Englewood Cliffs, N.J.; Figure 13: Rauck M. (1983) *Karl Freiherr Drais von Sauerbronn: Erfinder und Unternehmer (1785–1851)* Franz Steiner Verlag, Stuttgart; Figure 14: © Crown Copyright; Figure 15: John Pinkerton; Figures 16 and 46: Mary Evans Picture Library; Figure 17a: *Scientific American*, March 6th 1869; Figure 17b: *Scientific American,* February 5th 1887; Figures 17 (except a, b) 18, 44, 45, 49, 52, 53 and 54: Sharp A. (1896) *Bicycles and Tricycles*, Longmans, Green and Co; Figure 20: Sport and General Picture Library; Figure 21: Bicycle Association of Great Britain Ltd; Figure 22: Falcon Cycles Ltd; Figure 23: Allsport / Tony Duffy; Figure 24: by Dick Hargreaves from the New Y*ork Times Magazine*, 10th August 1980; Figure 25: Ballantine R. (1988) *Richard's New Bicycle Book*, Oxford Illustrated Press, © Richard Ballantine 1988; Figure 28: L*ook 'Cycle 1991' Catalogue*, Peugeot UK Ltd; Figure 30: A.J. Rogers and Partners; Figure 31: N*ew Civil Engineer*, 21st January 1988; Figure 32 (left): Ove Arup and Partners; Figure 33: Edison National Historic Site, National Park Service, U.S. Department of the Interior; Figure 34: U.S. Patent 2,297,691, Oct 6th 1942 (C.F. Carlson, Electrophotography); Figure 43: Andrew Ritchie © Photography 1982; Figure 47a, c, d: from Wilson S.S. 'Bicycle technology', Scientific American, vol. 228, no. 3, pp. 81–91, copyright © 1973 by Scientific American, Inc, all rights reserved; Figures 48 and 60: Museum of British Road Transport; Figures 50 and 51: Starley J.K.(1898) 'The evolution of the cycle', Journal of the Society of Arts, vol. 46, 20th May pp. 601–616; Figure 56: Science Museum, London; Figure 70: Courtesy of Eureka magazine, Findlay Publications Ltd; Figure 71: Middendorf W.H. (1969) 'Heat pump valve', Engineering Design, Allyn and Bacon, © Middendorf W.H., University of Cincinnati; Figure 73: Louis Psihoyos / Matrix; Figure 74: Gross A.C., Kyle C.R. and Malewicki D.J. (1983) Scientific American, December 1983; Figure 76: Flevo-Bike; Figure 77: Miles Kingsbury; Figure 78: The Hulton-Deutsch Collection; Figures 79, 80 and 82: Tony Hadland and Moulton Developments Ltd; Figure 81: © Crown Copyright; Figures 86, 88, and 91: Raleigh Industries Ltd; Figure 87: Dawes Cycles Ltd; Figure 90: Alex Moulton Ltd; Figure 92b: Bickerton Rawlinson Ltd; Figure 92c: Brompton Bicycle Ltd; Figure 92d: The Design Council; Figure 93 (top): © Crown Copyright; Figures 94 and 95 (bottom): Courtesy of J. Nesbit Evans and Co Ltd, Wednesbury; Figure 96: Science Museum, London; Figure 97: Mark Sanders; Figure 98: Jones D.B., 1989; Figure 99: Mines A., 1989; Figure 103: © Crown Copyright; Figure 104: Ray Thomas; Figure 105: Microwriter Systems; Figure 106: Times Newspapers Ltd; Figure 108: Pressens Bild AB, photo Olle Seijbold.

T264 DESIGN: PRINCIPLES AND PRACTICE

COURSE TEAM

Academics
Godfrey Boyle
Catherine Cooke
Nigel Cross
James Forster
Georgy Leslie
Ed Rhodes
Joe Rooney
Robin Roy
George Rzevski
Philip Steadman
David Walker

External Assessor Michael Tovey (Coventry Polytechnic)

Consultants
Michael Baker
Stephen Brown
Barry Dagger
Susan Hart
Jim Platts

BBC Producers
Cameron Balbirnie
Ian Spratley
Bill Young

Course Manager Ernie Taylor

Editors
Keith Cavanagh
Garry Hammond
Rodney Wilson

Graphic Designer Rob Williams

Graphic Artist Keith Howard

Media Librarian Caryl Hunter-Brown

Secretaries
Carole Marshall
Margaret Barnes
Jennie Conlon
Pat Dendy